Introduction to Animal Breeding

Introduction to Animal Breeding

David Thompson
Editor

KOROS PRESS LIMITED
London, UK

Introduction to Animal Breeding

© 2012

Printed in 2017 for Sale in the Indian Subcontinent

Published by
Koros Press Limited
3 The Pines, Rubery B45 9FF, Rednal,
Birmingham, United Kingdom

Tel.: +44-7826-930152
Email: info@korospress.com
www.korospress.com

ISBN: 978-1-78163-014-3

Editor: David Thompson

10 9 8 7 6 5 4 3 2 1

Printed in UK
British Library Cataloguing in Publication Data
A CIP record for this book is available from the British Library

Exclusively distributed by CBS Publishers & Distributors Pvt. Ltd.
Sales & Distribution Rights only for India, Pakistan, Bangladesh, Sri Lanka, Nepal and Bhutan.This book is not to be sold outside these territories.

Contents

Herbaceous Plants and Woody Plants • Use of Trade Mark for Animal Products • Design and Optimisation of Animal Breeding

Preface

The crossing of closely related animals is called Inbreeding. If this inbreeding is repeated continuously, it is called Upgrading. Inbreeding is used to retain desirable genetic traits in animals. Inbreeding is required in order to retain as many traits as possible by keeping the combination of genes intact. However, the inbreeding may result in homozygous recessive genes coming together to express some harmful phenotypic traits. Many breeding scientists have observed that hybrid vigour and fertility were lost due to repeated inbreeding. Such recessive and harmful genes are removed by some special techniques without sacrificing the major quality of the animal. If the race is relatively free of such harmful recessive genes, the process of inbreeding is a safe method for improvement of animals.

The crossing of distantly related animals is called Outbreeding. One of the problems the animal breeder faces in outbreeding is introduction of new genes into population. In this method it is possible to breed a desirable type of animal with a less desirable type and then to increase the degree of desirable traits. New and high yielding genes can be introduced into the population through outbreeding. In many cases these genes may come from a variety of stock. Out breeding in animals is useful for different purposes viz. To produce some valuable traits ii. To create new breeds iii. To produce a hybrid of superior vigour and value. To produce some valuable traits: Beef cattle may be crossed with dairy cattle to produce calves for superior veal (flesh) production. To create new breeds: A new breed is produced with desired characters from the two original breeds. This process of producing new breeds takes time. The present day breeds of animals have been developed through hybridisation.

A Mule is produced by crossing Equus equus (mare or female horse) and Equus homonius (jack or male ass). Mules are superior to horses in strength, endurance, resistance to disease and ability to work under unfavorable conditions. When a female mule is crossed with a jack, a colt is produced. The new traits into populations can also be induced through mutations. Since most of the mutations are harmful and the process of induction of mutations is quite expensive, this method of improvement of animals is impractical. It has been reported that a sheep in New England mutated in the direction of having shorter legs (a desirable quality) and formed the basis for racial improvement of sheep.

This book is essential reading for practicing researcher, students and scholars of subject and also the informed/interested general reader.

—*Editor*

In-Breeding and other Breeding Methods

One of the most hotly talked about topics with regard to pure-bred dogs is the use of in-breeding. This is a term that is often misused and is extremely misunderstood.

Part of the misunderstandings come from differences in the way the terms are used within the scientific/medical field, and how it is commonly used by breeders. These are the most commonly accepted definitions used by serious dog breeders and will be the definitions used within this article.

In-breeding-This is the breeding of closely related animals. Brother-Sister, Parent-Offspring, ½ brother-½ Sister.

Line-breeding-This is the breeding of animals that share common ancestors but are not closely related. For example the dogs may share a common great-grandparent.

Out-cross-This is generally considered the breeding of animals with no common ancestors within the first 4 or 5 generations.

Common Misconceptions

In-breeding causes genetic diseases-Breeding closely related animals increases the possibility that any bad genes in a line will show up. It does not 'cause' genetic disease.

Out-crossed dogs are healthier-This is only partly true. There is a known phenomenon called Hybrid Vigor. Two animals of unrelated strains breed and the offspring is often bigger and grows faster than it's purebred cousins. This method is often used by farmers in order to get their animals to market sooner. But one of the biggest misconceptions of hybrid vigour is that it applies to all animals of mixed heritage. Hybrid Vigour only applies to the animals that are the direct offspring of the crossing of the unrelated strains. In other words if you continue to breed animals of different strains there generally will not be any additional increase in hybrid vigour. If the unrelated strains share common genes for genetic disorders, hybrid vigour will not over ride the risk of the disorder showing up. Out-crossing can also cause problems if widely divergent physical types are mixed due to differences in growth rates and bone and muscle sizes.

Benefits of Each Type of Breeding

By definition, purebred dogs have a smaller gene pool to draw on than mixed breed dogs. That smaller gene pool gives the breed its individual characteristics, such as physical appearance and temperament. It is what makes a poodle a Poodle and a Golden Retriever a Golden Retriever. But there is considerable controversy with regard to whether the gene pools of the modern pure-bred have become too small.

Inbreeding

In-breeding is more likely to help "set" or "fix" a particular trait within a breed or a line by narrowing the gene pool to favour those traits. So if a breeder is looking to set a particular desirable feature of their line then in-breeding and choosing the offspring most strongly possessing that trait can be beneficial.

In the first pedigree Sydney is the maternal son of Annie, thus doubling on Annie's genes. Or another way to look at it is Annie is both Sydney's mother and Grand-mother. In the second example we have a full brother/sister in-breeding so we are doubling up on the genes from both Kiley and Annie.

In-breeding can also help identify those bad genes that exist within a line. Dogs possessing the bad genes can be eliminated from a breeding program and carriers also identified.

Intermittent in-breeding within a line or breed is not damaging to the long term health of the animals. However, in-breeding over successive generations can lead to reduced fitness and fertility problems among the offspring, resulting in a phenomena known as In-breeding Depression. It can take many generations to show up depending on the traits involved.

To use this method responsibly a breeder would not want to in-breed on animals with known genetic disorders, temperaments not in keeping with it's given breed, or known serious structural faults, or to in-breed frequently even on healthy-superior specimens.

Line-Breeding

Line-breeding is another way to help "set" or "fix" desirable traits. With line-breeding you breed animals that are related, but you are also routinely introducing genes from other lines into the genetic mix. It takes longer to fix the desirable traits this way, but doing so lowers the risk of those problems associated with repeated in-breeding. With a tight line-breeding you might find the same 3, 4 or more dogs showing up numerous times in a 5 generation pedigree.

Here's an example of a tightly line-bred pedigree (I've identified those dogs whose names show up more than once by colour)

Loose line-breeding over successive generations will result in more variations of physical appearance than would in-breeding or tight line-breeding, but will keep the physical look and structure within the same general size and shape, it also carries fewer long term risks.

And a sample of loose line-breeding (as with the above pedigree I'll identify dogs whose names repeat by colour):

According to geneticists. Line-breeding can be carried on for many many generations without deleterious effects on the line or breed as long as the individuals involved have few hidden genetic disorders.

Out-Crossing

Out-crossing in terms of pure-bred dogs is the breeding of unrelated dogs. On a pedigree no names will be repeated within the first 5 generations.

This type of breeding has both advantages and disadvantages. Which as it turns out are flip sides of the same argument. With out-crossing you are maintaining the greatest genetic diversity, but this also leads to the least consistency in terms of physical appearance and other traits.

Out-crossing does not guarantee that the animals won't develop genetic disorders, but it does tend to reduce the numbers of affected offspring. Your best chance of getting an animal that is less prone to developing a genetic disorder comes more from finding a conscientious breeder that screens their animals for hereditary disorders and breeds for the betterment of the breed.

All three methods of breeding have their place in a long term, well thought out breeding program. Talk to the breeder, ask questions as to what their goal is in doing a particular breeding. Ask about the risks and what problems are known to that line. And all lines have some because the perfect dog and the perfect lines are still goals of the future for all breeders.

Selective Breeding

This article is about selective breeding in the context of colloquial, everyday usage. For a more detailed treatment within the context of biology.

A Belgian Blue cow. The defect in the breed's myostatin gene is maintained through linebreeding and is responsible for its accelerated lean muscle growth.

Selective breeding is the process of breeding plants and animals for particular genetic traits. Typically, strains that are selectively bred are domesticated, and the breeding is sometimes done by a professional breeder. Bred animals are known as breeds, while bred plants are known as varieties, cultigens, or cultivars. The cross of animals results in what is called a crossbreed, and crossbred plants are called hybrids. The term selective breeding is synonymous with artificial selection.

In animal breeding techniques such as inbreeding, linebreeding, and outcrossing are utilized. In plant breeding, similar methods are used. Charles Darwin discussed how selective breeding had been successful in producing change over time in his book, *Origin of Species*. The first chapter of the book discusses selective breeding and domestication of such animals as pigeons, dogs and cattle. Selective breeding was used by Darwin as a springboard to introduce the theory of natural selection, and to support it.

Animal Breeding

Animals with homogeneous appearance, behaviour, and other characteristics are known as particular breeds, and they are bred through culling particular traits and selecting for others. Purebred animals have a single, recognizable breed, and purebreds with recorded lineage are called pedigreed. Crossbreeds are a mix of two purebreds, whereas mixed breeds are a mix of several breeds, often unknown. Animal breeding begins with breeding stock, a group of animals used for the purpose of planned breeding.

When individuals are looking to breed animals, they look for certain valuable traits in purebred stock for a certain purpose, or may intend to use some type of crossbreeding to produce a new type of stock with different, and, it is presumed, superior abilities in a given area of endeavour. For example, to breed chickens, a typical breeder intends to receive eggs, meat, and new, young birds for further reproduction. Thus, the breeder has to study different breeds and types of chickens and analyse what can be expected from a certain set of characteristics before he or she starts breeding them. Therefore, when purchasing initial breeding stock, the breeder seeks a group of birds that will most closely fit the purpose intended.

Purebred breeding aims to establish and maintain stable traits, that animals will pass to the next generation. By "breeding the best to the best," employing a certain degree of inbreeding, considerable culling, and selection for "superior" qualities, one could develop a bloodline superior in certain respects to the original base stock. Such animals can be recorded with a breed registry, the organization that maintains pedigrees and/or stud books. However, single-trait breeding, breeding for only one trait over all others, can be problematic. In one case mentioned by animal behaviourist Temple Grandin, roosters bred for fast growth or heavy muscles did not know how to perform typical rooster courtship dances, which alienated the roosters from hens and led the roosters to kill the hens after reproducing with them.

The observable phenomenon of hybrid vigour stands in contrast to the notion of breed purity. However, on the other hand, indiscriminate breeding of crossbred or hybrid animals may also result in degradation of quality.

Plant Breeding

Researchers at the USDA have selectively bred carrots with a variety of colors.

Plant breeding has been used for thousands of years, and began with the domestication of wild plants into uniform and predictable agricultural cultigens. High-yielding varieties have been particularly important in agriculture.

Selective plant breeding is also used in research to produce transgenic animals that breed "true" (i.e., are homozygous) for artificially inserted or deleted genes.

Selective Breeding in Aquaculture

Selective breeding in aquaculture holds high potential for the genetic improvement of fish and shellfish. Unlike terrestrial livestock, the potential benefits of selective breeding in aquaculture were not realized until recently. This is because high mortality led to the selection of only a few broodstock, causing inbreeding depression, which then forced the use of wild broodstock. This was evident in selective breeding programs for growth rate, which resulted in slow growth and high mortality.

Control of the reproduction cycle was one of the main reasons as it is a requisite for selective breeding programmes. Artificial reproduction was not achieved because of the difficulties in hatching or feeding some farmed species such as eel and yellowtail farming. A suspected reason associated with the late realisation of success in selective breeding programs in aquaculture was the education of the concerned people – researchers, advisory personnel and fish farmers. The education of fish biologists paid less attention to quantitative genetics and breeding plans.

Another was the failure of documentation of the genetic gains in successive generations. This in turn led to failure in quantifying economic benefits that successful selective breeding programs produce. Documentation of the genetic changes was considered important as they help in fine tuning further selection schemes.

Quality Traits in Aquaculture

Aquaculture species are reared for particular traits such as growth rate, survival rate, meat quality, resistance to diseases, age at sexual maturation, fecundity, shell traits like shell size, shell colour, etc.

Growth rate – growth rate is normally measured as either body weight or body length. (Gjedrem 1985). This trait is of great economic importance for all aquaculture species as faster growth rate speeds up the turnover of production (Gjedrem 1983). Improved growth rates show that farmed animals utilize their feed more efficiently through a correlated response (Gjedrem 1985).

Survival rate – survival rate may take into account the degrees of resistance to diseases (Gjedrem 1985). This may also see the stress response as fish under stress are highly vulnerable to diseases (Gjedrem 1983). The stress fish experience could be of biological, chemical or environmental influence.

Meat quality – the quality of fish is of great economic importance in the market. Fish quality usually takes into account size, meatiness, and percentage of fat, colour of flesh, taste, shape of the body, ideal oil and omega-3 content (Gjedrem 1985).

Age at sexual maturation-The age of maturity in aquaculture species is another very important for farmers as during early maturation the species divert all their energy to gonad production affecting growth and meat production and are more susceptible to health problems (Gjerde 1986).

Fecundity – As the fecundity in fish and shellfish is usually high it is not considered as a major trait for improvement. However, selective breeding practices may consider the size of the egg and correlate it with survival and early growth rate (Gjedrem 1985).

Finfish Response to Selection

Salmonids

Gjedrem (1979) showed that selection of Atlantic salmon (*Salmo salar*) led to an increase in body weight by 30% per generation. A comparative study on the performance of select Atlantic salmon with wild fish was conducted by AKVAFORSK Genetics Centre in Norway. The traits, for which the selection was done included growth rate, feed consumption, protein retention, energy retention, and feed conversion efficiency. Selected fish had a twice better growth rate, a 40% higher feed intake, and an increased protein and energy retention. This led to an overall 20% better Fed Conversion Efficiency as compared to the wild stock. Atlantic salmon have also been selected for resistance to bacterial and viral diseases. Selection was done to check resistance to Infectious Pancreatic Necrosis Virus (IPNV). The results showed 66.6% mortality for low-resistant species whereas the high-resistant species showed 29.3% mortality compared to wild species.

Rainbow trout (*S. gairdneri*) was reported to show large improvements in growth rate after 7-10 generations of selection. Kincaid et al. (1977) showed that growth gains by 30% could be achieved by selectively breeding rainbow trout for three generations. A 7% increase in growth was recorded per generation for rainbow trout by Kause et al. (2005). In Japan, high resistance to IPNV in rainbow trout has been achieved by selectively breeding the stock. Resistant strains were found to have an average mortality of 4.3% whereas 96.1% mortality was observed in a highly sensitive strain. Coho salmon (*Oncorhynchus kisutch*) increase in weight was found to be more than 60% after four generations of selective breeding. In Chile, Neira et al. (2006) conducted experiments on early spawning dates in coho salmon. After selectively breeding the fish for four generation, spawning dates were 13 – 15 days earlier.

Cyprinids

Selective breeding programs for the Common carp (*Cyprinus carpio*) include improvement in growth, shape and resistance to disease. Experiments carried out in the USSR used crossings of broodstocks to increase genetic diversity and then selected the species for traits like growth rate, exterior traits and viability, and/or adaptation to environmental conditions like variations in temperature. Kirpichnikov et al. (1974) and Babouchkine (1987) selected carp for fast growth and tolerance to cold, the Ropsha carp. The results showed a 30-40% to 77.4% improvement of cold tolerance but did not provide any data for growth rate. An increase in growth rate was observed in the second generation in Vietnam. Moav and Wohlfarth (1976) showed positive results when selecting for slower growth for three generations compared to selecting for faster growth. Schaperclaus (1962) showed resistance to the dropsy disease wherein selected lines suffered low mortality (11.5%) compared to unselected (57%).

Channel Catfish

Growth was seen to increase by 12 – 20% in selectively bred Iictalurus punctatus (Bondari, 1983). More recently, the overall response of Channel Catfish response to selection for improved growth rate was found to be approximately 80%, i.e., an average of 13% per generation (Dunham 2006).

Shellfish Response To Selection

Oysters

Selection for live weight of Pacific oysters showed improvements ranging from 0.4% to 25.6% compared to the wild stock. Sydney-rock oysters (*Saccostrea commercialis*) showed a 4% increase after one generation and a 15% increase after two generations. Chilean oysters (*Ostrea chilensis*), selected for improvement in live weight and shell length showed a 10-13% gain in one generation. Bonamia ostrea is a protistan parasite that causes catastrophic losses (nearly 98%) in European flat oyster *Ostrea edulis* L. This protistan parasite is endemic to three oyster-regions in Europe. Selective breeding programs show that *O. edulis* susceptibility to the infection differs across oyster strains in Europe. A study carried out by Culloty et al. (2001) showed that 'Rossmore' oysters in Cork harbour, Ireland had better resistance compared to other Irish strains. A selective breeding program at Cork harbour uses broodstock from 3– to 4-year-old survivors and is further controlled until a viable percentage reaches market size. Over the years 'Rossmore' oysters have shown to develop lower prevalence to *B. ostreae* infection and percentage mortality. Ragone Calvo et al. (2003) selectively bred the eastern oyster, *Crassostrea virginica*, for resistance against co-occurring parasites *Haplosporidium nelson* (MSX) and *Perkinsus marinus* (Dermo). They achieved dual resistance to the disease in four generations of artificial selection. The oysters showed higher growth and survival rates and low susceptibility to the infections. At the end of the experiment, artificially selected *C. virginica* showed a 34-48% higher survival rate.

Penaeid Shrimps

Selection for growth in Penaeid shrimps yielded successful results. A selective breeding program for *Litopenaeus stylirostris* saw an 18% increase in growth after the fourth generation and 21% growth after the fifth generation (Goyard et al. 1999). *Marsupenaeus japonicas* showed a 10.7% increase in growth after the first generation. Argue et al. (2002) conducted a selective breeding program on the Pacific White Shrimp, *Litopenaeus vannamei* at The Oceanic Institute, Waimanalo, USA from 1995 to 1998.

They reported significant responses to selection compared to the unselected control shrimps. After one generation, a 21% increase was observed in growth and 18.4% increase in survival to TSV. The Taura Syndrome Virus (TSV) causes mortalities of 70% or more in shrimps. C.I. Oceanos S.A. in Colombia selected the survivors of the disease form infected ponds and used them as parents for the next generation. They achieved satisfying results in two or three generations wherein survival rates approached levels before the outbreak of the disease (Cock et al. 2009). The resulting heavy losses (up to 90%) caused by Infectious hypodermal and haematopoietic necrosis virus (IHHNV) caused a number of shrimp farming industries started to selectively breed shrimps resistant to this disease. Successful outcomes led to development of Super Shrimp, a selected line of *L. stylirostris* that is resistant to IHHNV infection. Tang et al. (2000) confirmed this by showing no mortalities in IHHNV-challenged Super Shrimp post larvae and juveniles.

Aquatic Species Versus Terrestrial Livestock

Selective breeding programs for aquatic species provide better outcomes compared to terrestrial livestock. This higher response to selection of aquatic farmed species can be attributed to the following:

High fecundity in both sexes fish and shellfish enabling higher selection intensity.

Large phenotypic and genetic variation in the selected traits.

Selective breeding in aquaculture provide remarkable economic benefits to the industry, the primary one being that it reduces production costs due to faster turnover rates. This is because of faster growth rates, decreased maintenance rates, increased energy and protein retention, and better feed efficiency. Applying such genetic improvement program to aquaculture species will increase productivity to meet the increasing demands of growing populations.

Artificial Selection

Artificial selection (or selective breeding) describes intentional breeding for certain traits, or combination of traits. The term was utilized by Charles Darwin in contrast to natural selection, in which the differential reproduction of organisms with certain traits is attributed to improved survival or reproductive ability ("Darwinian fitness"). As opposed to artificial selection, in which humans favour specific traits, in natural selection the environment acts as a sieve through which only certain variations can pass.

The deliberate exploitation of artificial selection has become very common in experimental biology, as well as the discovery and invention of new drugs.

Artificial selection can also be unintentional; it is thought that domestication of crops by early humans was largely unintentional.

Historical Development

Artificial selection was practiced by the Romans. Treatises as much as 2,000 years old give advice on selecting animals for different purposes, and these ancient works cite still older authorities, such as Mago the Carthaginian. The notion of artificial selection was later expressed by the Persian Muslim polymath Abu Rayhan Biruni in the 11th century. He noted the idea in his book titled India, *and gave various examples.* The agriculturist selects his corn, letting grow as much as he requires, and tearing out the remainder. The forester leaves those branches which he perceives to be excellent, whilst he cuts away all others. The bees kill those of their kind who only eat, but do not work in their beehive.

Charles Darwin coined the term as an illustration of his proposed wider process of natural selection. Darwin noted that many domesticated animals and plants had special properties that were developed by intentional animal and plant breeding from individuals that showed desirable characteristics, and discouraging the breeding of individuals with less desirable characteristics.

Slow though the process of selection may be, if feeble man can do much by his powers of artificial selection, I can see no limit to the amount of change, to the beauty and infinite complexity of the co-adaptations between all organic beings, one with another and with their physical conditions of life, which may be effected in the long course of time by nature's power of selection. We are profoundly ignorant of the causes producing slight and unimportant variations; and we are immediately made conscious of this by reflecting on the differences in the breeds of our domesticated animals in different countries,—more especially in the less civilized countries where there has been but little artificial selection.

Contrast to Natural Selection

There is no real difference in the genetic processes underlying artificial and natural selection, and the concept of artificial selection was used by Charles Darwin as an illustration of the wider process of natural selection. The selection process is termed "artificial" when human preferences or influences have a significant effect on the evolution of a particular population or species. Indeed, many evolutionary biologists view domestication as a type of natural selection and adaptive change that occurs as organisms are brought under the control of human beings.

However, it is useful to distinguish between artificial selection that is unintentional or involves manipulating the environment only, and artificial selection that alter internal DNA sequences in the laboratory. Genetic manipulation in the labs have little in common to processes that occur in nature.

Laboratory Usage

The deliberate exploitation of selective power has become common in experimental biology, particularly in microbiology and genetics. In a ubiquitous laboratory technique in genetic engineering, genes are introduced into cells in cell culture, usually bacteria, on a small circular DNA molecule called a plasmid in a process called transfection. The gene of interest is accompanied on the plasmid by a reporter gene, or "selectable marker", which encodes a specific trait such as antibiotic resistance or ability to grow in high salt concentrations. The cells can then be cultured in an environment that would kill normal cells, but is hospitable to those that have taken up and expressed the genes on the plasmid. In this way expression of the reporter gene serves as a signal that the gene of interest is also being expressed in the cells. Another technique used in drug development uses an iterative selective process called in vitro selection to evolve aptamers, or nucleic acid fragments capable of binding specific organic compounds with high binding affinity. Studies in evolutionary physiology, behavioural genetics, and other areas of organismal biology have also made use of deliberate artificial selection, though longer generation times and greater difficulty in breeding can make such projects challenging in vertebrates.

Breed Registry

A breed registry, also known as a stud book or register, in animal husbandry and the hobby of animal fancy, is an official list of animals within a specific breed whose parents are known. Animals are usually registered by their breeders when they are still young. The terms "stud book" and "register" are also used to refer to lists of male animals "standing at stud", that is, those animals actively breeding, as opposed to every known specimen of that breed. Such registries usually issue certificates for each recorded animal, called a Pedigree, Pedigreed animal documentation, or most commonly, an animal's "papers". Registration papers may consist of a simple certificate or a listing of ancestors in the animal's background, sometimes with a chart showing the lineage.

Types of Registries

There are breed registries and breed clubs for several species of animal, such as dogs, horses, cows and cats. The Association of Zoos and Aquariums (AZA) also maintains stud books for captive species on display ranging from aardvarks to zebras.

Kennel clubs always maintain registries, either directly or through affiliated dog breed clubs. Some multi-breed clubs also maintain registries, as do non-affiliated breed clubs, and there are a few registries that are maintained by other private entities such as insurance agencies; an example of this in the United States is the Field Dog Stud Book. Working dog organizations also maintain registries.

There are also entities which refer to themselves as registries, but which are thinly veiled marketing devices for vendors of puppies and adult dogs, as well as a means of collecting registration fees from novice dog owners unfamiliar with reputable registries and breed clubs. Though these entities generally focus on dogs, particularly in relationship to the puppy mill industry, some are marketed as cat registries. At least one group claims to register wild species (held by private individuals rather than by legitimate zoological parks, which use the AZA).

Horse breeding also has such problematic registries, particularly for certain colour breeds. While many colour breeds are legitimate, some "registries" are primarily a marketing tool for poor quality animals that are not accepted for registration by more mainstream organizations. Other "registries" are marketing attempts to create new horse breeds, usually by breeders using crossbreeding to create a new type, but the animals are not yet breeding "true."

Many such questionable registries are incorporated as for-profit commercial businesses, in contrast to the formal not-for-profit status of most reputable breed clubs. They may provide volume discounts for registrations by commercial dog breeders such as puppy mills. An unscrupulous registry for dogs or horses is often spotted by a policy to not require any proof of pedigree at all. In the dog world, such registries may not sponsor competitions, and thus cannot award championship points to identify the best individuals registered within a particular breed or species. In the less-organized world of horse shows, where many different sanctioning organizations exist, some groups sponsor their own competitions, though wins at such events seldom carry much prestige in mainstream circles.

Some registers have the word "registry" in their title used in the sense of "list"; these entities are not registers in the usual sense in that they do not maintain breeding records. In the dog world, listed animals are required to be de-sexed. The American Mixed Breed Obedience Registry is an example. Some equestrian organizations create a recording system for tracking the competition records of horses, but, though horses of any sex may be recorded, they also do not maintain breeding or progeny records. The United States Equestrian Federation is one organization that uses such a system.

Closed

A closed stud book is a stud book or breed registry that does not accept any outside blood. The registered animals and all subsequent offspring trace back to the foundation stock. This ensures that the animal is a purebred member of the breed. In horses, an example of a closed stud book is that of the Thoroughbred, with a stud book tracing to 1791. The American Kennel Club is an example of a kennel club with primarily closed books for dogs; it allows new breeds to develop under its Foundation Stock Service, but such dogs are not eligible for competition in AKC conformation shows. For the breed to move to the Miscellaneous class and then to fully recognized status, the breed's stud book must be closed.

A closed stud book allows the breed to stay very pure to its type, but limits its ability to be improved. This may put a breed at a disadvantage, especially in performance disciplines, where an animal is worth more if it is successful in competition even if it is not pure. It also limits the gene pool, which may make certain undesirable characteristics become accentuated in the breed, such as a poor conformational fault or a disease. It also, depending on original numbers and management practices, can lead to an ever increasing level of inbreeding.

Some closed stud books, particularly for certain European breeds such as the Finnhorse and the Trakehner, may also have a studbook selection criteria where animals must meet either a conformation standard, a performance standard, or both.

Open

In an open stud book, animals may be registered even if their parents or earlier ancestors were not previously registered with that particular entity. Usually an open stud book has strict studbook selection criteria that require an animal to meet a certain standard of conformation, performance or both. This allows breeders to modify breeds by including individuals who conform to the breed standard but are of outside origin. Some horse breeds allow crossbreds who meet specific criteria to be registered. One example is the semi-open stud book of the American Quarter Horse, which still accepts horses of Thoroughbred breeding, particularly via its appendix registry. Among dogs, an example of an open stud book would be the registries maintained by the American Kennel Club as its Foundation Stock Service. The SACBR uses an open stud book system to register all purebred dogs with or without ancestry. In some cases, an open stud book may eventually become closed once the breed type is deemed to be fully set.

In some agricultural breeds, an otherwise closed registry includes a *grading up* route for the incorporation of cross-bred animals. Often such incorporation is limited to females, with the progeny only being accepted as full pedigree animals after several generations of breeding to full-blood males. Such mechanisms may also allow the incorporation of purebred animals descended from unregistered stock or of uncertain parentage.

More controversial open stud books are those where there are few, if any qualifications for animals other than a single trait, such as a "colour breed," particularly when the colour is not a true-breeding characteristic. However, some breeds have a standard colour or colour preference that is one criterion among others used to register animals.

Appendix Registries

Some open or partly open registries may permit animals who have some but not all qualifications for full registration to nonetheless be entered in a preliminary recording system often called an "appendix" registry.

The most notable is that of the American Quarter Horse Association, which allows part-Thoroughbred/part-Quarter Horse foals to be recorded and shown, with full registration allowed after the horse achieves a set performance or merit standard akin to that of a merit registry. Other appendix registries are seen in certain colour breeds of horses, such as the Appaloosa, American Paint Horse, and American Cream Draft Horse, where foals with the proper pedigree for registration but do not meet the colour standard for the breed, yet may still carry the necessary genetics in a minimally-expressed form, may be registered and bred to fully registered animals, with ensuing offspring eligible for registration if they meet the breed standard.

Performance or Merit

Another form of open registry is a registry based on performance or conformation, called in some societies *Registry on Merit*. In such registries, an eligible animal that meets certain criteria is eligible to be registered *on merit*, regardless of ancestry. In some cases, even unknown or undocumented ancestry may be permitted.

The Registry on Merit or ROM may be tied to percentage of bloodline, conformation, or classification or may be based solely on performance.

In the horse world, many Warmblood breeds require a conformation and performance standard for registration, and often allow horses of many different breeds to qualify, though documented pedigrees are usually required. Some breed registries use a form of ROM in which horses at certain shows may be sight classified. For example, at qualifying shows in Australia, winning horses of stock-type breeding receive points for conformation, which are attested to by the judges and recorded in an owner's special book. The points are accumulated to eventually result in a Registry on Merit.

Registry on Merit is prevalent with sheepdog registries, in particular those of the Border Collie, and some other breeds with a heavy emphasis on working ability. In this type of ROM, the dog's conformation and ancestry generally does not matter.

Papers

Breed registries usually issue certificates for each recorded animal, called a Pedigree, Pedigreed animal documentation, or most commonly, an animal's "papers". Registration papers may consist of a simple certificate or a listing of ancestors in the animal's background, sometimes with a chart showing the lineage. Usually, there is space for the listing of successive owners, who must sign and date the document if the animal is gifted, leased or sold. Papers transferred upon sale of an animal may be submitted to the registry in order to update the ownership information, and in most cases, the registry will then issue a new set of papers listing the new owner as the proper owner of the horse. Genuine papers are often identifiable as containing the registered name and number of the individual animal and its date of birth, the name of the attesting organization, with the logo if there is one, the name and signature of the registrar or other authorized person, and a corporate stamp or seal.

Documentation usually included on registration certificates or papers includes:

- name of sire (father) and dam (mother)
- names of other ancestors, to the number of generations required by the issuing organization
- In dogs, details of the litter this animal came from
- its colour and markings
- name, address and registered number of the breeder (often defined as the owner of the female at the time of the animal's conception or birth)
- name and address of the original owner who registered the foal.

Crossbreeding and Backbreeding

In some registries, breeders may apply for permission to crossbreed other breeds into the line to emphasize certain traits, to keep the breed from extinction or to alleviate problems caused in the breed by inbreeding from a limited set of animals. A related preservation method is backbreeding, used by some equine and canine registries, in which crossbred individuals are mated back to purebreds to eliminate undesirable traits acquired through the crossbreeding.

Registered Names and Naming Traditions

Naming rules vary according to the species and breed being registered. For example, show horses have a registered name, that is, the name under which they are registered as a purebred with the appropriate breed registry, and purebred dogs intended for the sport of conformation showing must be registered with the kennel club in which they will compete; and although there are no specific naming requirements, there are many traditions that may be observed in naming.

Along with a registered name, these animals often also have a simpler "pet name" known as a call name for dogs or a stable name for horses, which is used by their owners or handlers when talking to the animal. For example, the famous Thoroughbred race horse Man o' War was known by his stable name, "Big Red." The name can be anything that the animal's owner prefers. For example, the dog that won the 2008 Westminster show (US) was named Ch K-Run's Park Me In First, with the call name of "Uno".

Dogs in the breed registry of a working dog club (particularly herding dogs) must usually have simple, no-nonsense monikers deemed to be "working dog names" such as "Pal," "Blackie," or "Ginger." The naming rules for independent dog clubs vary but are usually similar to those of kennel clubs.

The registered name often refers directly or indirectly to the breeder of the animal. Traditionally, the breeder's kennel prefix form the first part of the dog's registered name. For example, all dogs bred at the Gold Mine Kennels would have names that begin with the words Gold Mine. Horse breeders are usually not required to do this, but often find it to be a good form of commercial promotion to include a stable name or farm initials in the horse's name. For example, Gold Mine Stables may name give all horses names with the prefix "Gold Mine," "GM," or "GMS." The Jockey Club, which registers Thoroughbreds, requires stable names to be registered, but does not require their use in animal names.

Many dog breeders name their puppies sequentially, based on litter identification: Groups of puppies may be organized as Litter A, Litter B, and so on. When this is done, the names of all the puppies in litter A start with the letter "A," then "B" for litter B and so on. Horse breeders, especially in Europe, sometimes use the first letter of the dam's name as the first letter in the name of all of her offspring. Other breeders may use the same first letter to designate all the foals born on the farm in a given year.

Some breeders create a name that incorporates or acknowledges the names of the sire, dam or other forebears. For example, the famous cutting horse Doc O'Lena was by Doc Bar out of Poco Lena, a daughter of Poco Bueno. Some names are a little less direct; 2003 Kentucky Derby winner Funny Cide was by Distorted Humor out of Belle's Good Cide, and the famous race horse Native Dancer was by Polynesian out of Geisha. Other breeders use themes. For example, a more imaginative breeder at the Gold Mine Kennels might name all the puppies of one litter after green precious stones: Gold Mine Emerald, Gold Mine Jade, and Gold Mine Peridot. Names for a subsequent litter might start with the adjectives describing precious stones: Gold Mine Sparkle, Gold Mine Brilliance, and Gold Mine Chatoyant. Breeders may be as creative or as mundane as they wish.

In order to minimize the unwieldiness that long and fancy names can bring, registries usually limit the total number of characters and sometimes number of separate words that may compose the animal's registered name. They are often prohibited from using only punctuation or odd capitalization to create a unique name; names are often published in all capitals on registration papers. Breeders are generally not allowed to use any name that may be obscene or misleading, such as the word 'champion' in a name, a trademark, or anything that can be mistaken for the name of another kennel or, sometimes, stable. Only after an animal has achieved a legitimate championship will some registries permit the use of the prefix Ch. or other title before or after their registered name. Some registries may use symbols to designate the status of certain individuals. An asterisk * may be used to designate an animal born in another country and imported. A plus + may be used to designate a champion or an animal under special registration status.

Cat Registry

A cat registry is an organisation that registers cats for exhibition and breeding purposes. A cat registry stores the pedigrees (genealogies) of cats, prefixes or affixes of catteries, studbooks (lists of authorised studs of recognised breeds), breed descriptions and the standards of points (SoP) for those breeds; lists of judges qualified to judge at shows run by, or affiliated with, that registry. A cat registry is not the same as a cat club or breed society (these may be affiliated with one or more registries with whom they have lodged breed standards in order to be able to exhibit under the auspices of that registry).

The first cat registry was the National Cat Club, set up in 1887 in England. Until the formation of the Governing Council of the Cat Fancy in 1910, the National Cat Club was also the Governing Body of the Cat Fancy. A rival registry called the Cat Club was set up in 1898, but foundered in 1903 and was replaced by the Cat Fanciers Association. Cats could only be registered with one or other registry. These two fancies merged in 1910 and became the GCCF. In the USA, the 1899 Chicago cat show resulted in the formation of the Chicago Cat Club, followed by the more powerful Beresford Cat Club (named after noted British breeder Lady Marcus Beresford). In 1906, the American Cat Association became the main registry. In 1908 this became the Cat Fanciers' Association Inc (CFA).

In the intervening years, many cat registries have been formed worldwide. These range from international organisations or federations to national registries in one particular country. In many countries, independent registries have also been formed which may or may not be recognised by the main registries. While some cat registries forbid the practice, it is now common to allow a cat to be registered by more than one registry. The largest overall organisation is the Fédération Internationale Féline (FIFe) which is a worldwide federation of member cat registries.

Cat Registries have their own rules and also organise or licence cat shows. The show procedures vary widely and awards won in one registry are not normally recognised by another.

The World Cat Congress (WCC) is an international coordinating organisation of the largest cat registries. The WCC operates an "open Doors" policy by which cats registered with one registry can be shown under the rules of another registry.

Some independent cat registries specialise in particular types of cat that are ineligible for registration with a major registry due to breed restrictions or certain genetic traits. For example The Dwarf Cat Association recognises breeds derived from the short-legged Munchkin (a cat body type genetic mutation) which are banned by FIFe and some other registries, while the Rare and Exotic Feline Registry specialises in cats derived from (or alleged to derive from) hybrids with wildcat species.

Recognition Levels

Most registries offer several levels of recognition (often called registers). The actual designations differ between registries, but typically these are:Full a breed that competes for championship titles at shows organised by, or affiliated to, that registry.

Provisional/Preliminary the level of recognition of cat breeds until they demonstrate that they breed true to their registered standards; there may be several levels of provisional/preliminary recognition e.g. *new* or *advanced* as numbers and popularity increase.

Experimental a provisional register for breeds in development; this may be separate from the provisional/preliminary register in some cat fancies.

Exhibition only a new trait, new import or minority variety that does not compete, but is exhibited in order to attract opinion and/or potential breeders.

Registration only status means cats of that breed can be registered, but do not have permission to be exhibited. Not all breeds achieve Full (championship) status. There may also be Active and Inactive registers that denote whether a cat may be legitimately used in breeding and its offspring registered. In breeds known to carry recessive genes (e.g. longhaired cats born from shorthaired parents, colourpointed cats born from non-colourpointed parents), cats that do not meet their breed standard might be registered as variants or they might be registered under a different breed name. These may sometimes be used to maintain a good gene pool, but not exhibited in Championship classes for the parents' breed.

A Genetic Register is used by some registries for breeds where a genetic test is required before cats can be bred from. Cats that have not been cleared through testing remain on the genetic register until negative test results are provided.

A cat registry is at liberty to refuse to accept breeds if it feels the breed is not genetically sound; does not breed true to the standard put forward by the developer(s) of the breed (with allowances made for known variants); is not represented in sufficient numbers or is not sufficiently distinct from breeds already recognised by the registry. It may also expel breeders who do not conform to accepted standards of behaviour and ethics, with the result that their cats may be disqualified from its shows.

The rules as to what constitutes a new breed vary from registry to registry. TICA is a relatively progressive registry that will recognise breeds derived from crossing existing breeds; mutations of an existing breed; naturally occurring breeds indigenous to a geographical location; a breed already recognised by a different registry and experimental breeds that do not yet have a TICA-approved breed name. FIFe will register some new breeds imported from other registries but have set procedures for these breeds to gain full recognition. The GCCF is a more conservative registry and recognise new colour variations of an existing breed, but do not usually recognise other mutations of an existing breed e.g. spontaneous rexed fur.

Breed Numbers, Acronyms and Codes

Registries allocate a breed number, acronym or a Code to the breeds they register. Most use a two or 3 letter acronym e.g. MK (Munchkin), JBT (Japanese Bobtail). This may be followed by numbers or lower case acronyms that indicate colour and pattern, these being subdivisions of the breed. For historical reasons, the British GCCF allocate numbers to breeds and the Black Persian Longhair is registered under a different breed number, and effectively as a different breed, to the Blue Persian Longhair. These lists may be found on individual registry websites (or in their printed publications where they do not yet have a website). All FIFe Member cat registries use the EMS (Easy Memory System) breed and variety code which consists of a breed abbreviation followed by pattern and colour letters and digits which are consistent across all breeds.

Where a breed is already recognised by another registry, it is becoming increasingly common to adopt an existing acronym (with the possible addition or subtraction of a letter) in order to avoid clashes and confusion. Where 2 breeds with different characteristics have the same name, it is usual to prefix the name with the country/area of origin e.g. in the US the "Burmese" and "European Burmese" are different breeds with different conformation. In the UK, "Burmese" refers to the European form as the "American Burmese" is not recognised.

A single breed may have 2 different breed names in different countries. In Britain, a cat of Persian type with the colourpoint pattern is called a Colourpoint Persian. In the USA it is called a Himalayan. The American-bred Serengeti was founded in 1992 by Karen Sausaman from Oriental x Bengal crosses to resemble the wild cats of the Serengeti plains but without the introduction of wild cat blood. In Britain, a Bengal x Siamese cross was originally called the Savannah, but was later renamed Serengeti because of an existing American breed called the Savannah. The American-bred Savannah resembles the Serval and the first generation cross is Serval x domestic.

Where colours have been added to a breed through outcrossing to another breed, not all registries accept the new colours under the original breed name e.g. Chocolate Persians and Lilac Persians may be recognised under the name "Kashmir" as the two colors were introduced through crossing to Siamese cats during the development of the Colourpoint Persian (UK) and Himalayan (USA).

Dog Breeding

Dog breeding is the practice of mating selected specimens with the intent to maintain or produce specific qualities and characteristics. A litter of puppies and their motherDogs reproduce without human interference, so their offsprings' characteristics are determined by natural selection. Domestic dogs may be intentionally bred by their owners. A person who intentionally mates dogs to produce puppies is referred to as a dog breeder. Breeding relies on the science of genetics, so the breeder with a knowledge of canine genetics, health, and the intended use for the dogs attempts to breed suitable dogs.

Humans have maintained populations of useful animals around their places of habitat since pre-historic times. They have intentionally fed dogs considered useful, while neglecting or killing others, thereby establishing a relationship between humans and certain types of dog over thousands of years. Over these millennia, domesticated dogs have developed into distinct types, or groups, such as livestock guardian dogs, hunting dogs, and sighthounds. To maintain these distinctions, humans have intentionally mated dogs with certain characteristics to encourage those characteristics in the offspring.

Through this process, hundreds of dog breeds have been developed. Initially, the ownership of working and, later, purebred dogs, was a privilege of the wealthy. Nowadays, many people can afford to buy a dog. Some breeders chose to breed purebred dogs, while some prefer to produce crossbred dogs, claiming that the outcross is healthier than original breeds, and avoiding linebreeding or inbreeding.

Registries

Breeders may report the birth of a litter of puppies to a dog registry, such as kennel club to record it in stud books such as those kept by the AKC (American Kennel Club). Such registries maintain records of dogs' lineage and are usually affiliated with kennel clubs. Maintaining correct data is important for purebred dog breeding. Access to records allows a breeder to analyse the pedigrees and anticipate traits and behaviors.

Requirements for the breeding of registered purebreds vary between breeds, countries, kennel clubs and registries. Breeders have to abide the rules of the specific organization to participate in its breed maintenance and development programs. The rules may apply to the health of the dogs, such as joint x-rays, hip certifications, and eye examinations; to working qualities, such as passing a special test or achieving at a trial; to general conformation, such as evaluation of a dog by a breed expert. However, many registries, particularly those in North America, are not policing agencies that exclude dogs of poor quality or health. Their main function is simply to register puppies born of parents who are themselves registered.

Criticism

The term 'backyard breeders' is commonly used in Canada and the U.S. to describe a breeder with a lack of knowledge and experience; while the term 'puppy mills' or 'puppy farms' refers to businesses that mass produce puppies of different breeds. Animal rights activists claim that breeding dogs in order to sell them is unethical, attacking breeders whom they believe are more concerned with profit than the animals' welfare. Critics cite breed registries for encouraging the inbreeding of dogs thereby contributing to a proliferation of genetic disorders.

Genetic Defects

Some dogs have certain inheritable characteristics that can develop into a disability or disease. Excessive wear of hip joint or bone, known as hip dysplasia is one such condition. As well, some eye abnormalities, heart conditions, deafness, are proven to be inherited. There have been extensive studies of these conditions, commonly sponsored by breed clubs and dog registries, while breed clubs provide information of common genetic defects for according breed. As well, special organizations, such as Orthopedic Foundation for Animals, collect data and provide it to breeders, as well as to the general public. Some registries, such as American Kennel Club include records of absence of certain genetic defects, known as certification, into dog's individual records. For example, the German Shepherd National Breed Club in Germany is a registry that recognizes that hip dysplasia is a genetic defect for the dogs of this breed. Accordingly, it requires all dogs to pass evaluation for absence of Hip Dysplasia in order to register their progeny, and records the results in individual dog's pedigrees.

Horse Breeding

Horse breeding is reproduction in horses, and particularly the human-directed process of selective breeding of animals, particularly purebred horses of a given breed. Planned matings can be used to produce specifically desired characteristics in domesticated horses. Furthermore, modern breeding management and technologies can increase the rate of conception, a healthy pregnancy, and successful foaling.

Terminology

The male parent of a horse, a stallion, is commonly known as the sire and the female parent, the mare, is called the dam. Both are genetically important, as each parent provides half of the genetic makeup of the ensuing offspring, called a foal. (Contrary to popular misuse, the word "colt" refers to a young male horse only; "filly" is a young female.) Though many horse owners may simply breed a family mare to a local stallion in order to produce a companion animal, most professional breeders use selective breeding to produce individuals of a given phenotype, or breed. Alternatively, a breeder could, using individuals of differing phenotypes, create a new breed with specific characteristics.

A horse is "bred" where it is foaled (born). Thus a foal conceived in England but foaled in the United States is regarded as being bred in the US. In some cases, most notably in the Thoroughbred breeding industry, American-bred horses may also be described by the state in which they are foaled. Some breeds denote the country, or state, where conception took place as the origin of the foal.

Similarly, the "breeder", is the person who owned or leased the mare at the time of foaling. That individual may not have had anything to do with the mating of the mare. It's important to review each breed registry's rules to determine which applies to any specific foal.

In the horse breeding industry, the term "half-brother" or "half-sister" only describes horses which have the same dam, but different sires. Horses with the same sire but different dams are simply said to be "by the same sire", and no sibling relationship is implied. "Full" (or "own") siblings have both the same dam and the same sire. The terms paternal half-sibling, and maternal half-sibling are also often used. Three-quarter siblings are horses out of the same dam, and are by sires that are either half-brothers (i.e. same dam) or who are by the same sire.

Thoroughbreds and Arabians are also classified through the "distaff" or direct female line, known as their "family" or "tail female" line, tracing back to their taproot foundation bloodstock or the beginning of their respective stud books. The female line of descent always appears at the bottom of a tabulated pedigree and is therefore often known as the bottom line.

"Linebreeding" technically is the duplication of fourth generation or more distant ancestors. However, the term is often used more loosely, describing horses with duplication of ancestors closer than the fourth generation. It also is sometimes used as a euphemism for the practice of inbreeding, a practice that is generally frowned upon by horse breeders, though used by some in an attempt to fix certain traits.

Estrous Cycle of the Mare

Stallion checking a mare in estrus. The mare welcomes the stallion by lowering her rear and lifting her tail. The estrous cycle (also spelled oestrous) controls when a mare is sexually receptive toward a stallion, and helps to physically prepare the mare for conception. It generally occurs during the spring and summer months, although some mares may be sexually receptive into the late fall, and is controlled by the photoperiod (length of the day), the cycle first triggered when the days begin to lengthen. The estrous cycle lasts about 19–22 days, with the average being 21 days. As the days shorten, the mare returns to a period when she is not sexually receptive, known as anestrus. Anestrus-occurring in the majority of, but not all, mares-prevents the mare from conceiving in the winter months, as that would result in her foaling during the harshest part of the year, a time when it would be most difficult for the foal to survive.

This cycle contains 2 phases:

Estrus, or Follicular, phase: 5–7 days in length, when the mare is sexually receptive to a stallion. Estrogen is secreted by the follicle. Ovulation occurs in the final 24–48 hours of estrus.

Diestrus, or Luteal, phase: 14–15 days in length, the mare is not sexually receptive to the stallion. The corpus luteum secretes progesterone.

Depending on breed, on average, 16% of mares have double ovulations, allowing them to twin, this does not affect the length of time of estrus or diestrus.

Effects on the Reproductive System During the Estrous Cycle

Changes in hormone levels can have great effects on the physical characteristics of the reproductive organs of the mare, thereby preparing, or preventing, her from conceiving.

Uterus: increased levels of estrogen during estrus cause edema within the uterus, making it feel heavier, and the uterus loses its tone. This edema decreases following ovulation, and the muscular tone increases. High levels of progesterone do not cause edema within the uterus. The uterus becomes flaccid during anestrus.

Cervix: the cervix starts to relax right before estrus occurs, with maximal relaxation around the time of ovulation. The secretions of the cervix increase. High progesterone levels (during diestrus) cause the cervix to close and become toned.

Vagina: the portion of the vagina near the cervix becomes engorged with blood right before estrus. The vagina becomes relaxed and secretions increase.

Vulva: relaxes right before estrus begins. Becomes dry, and closes more tightly, during diestrus.

Hormones Involved in the Estrous Cycle, During Foaling, and After Birth

The cycle is controlled by several hormones which regulate the estrous cycle, the mare's behaviour, and the reproductive system of the mare. The cycle begins when the increased day length causes the pineal gland to reduce the levels of melatonin, thereby allowing the hypothalamus to secrete GnRH.

GnRH (Gonadotropin releasing hormone): secreted by the hypothalamus, causes the pituitary to release of 2 gonadotrophins: LH and FSH.

LH (Luteinizing hormone): levels are highest 2 days following ovulation, then slowly decrease over 4–5 days, dipping to their lowest levels 5–16 days after ovulation. Stimulates the follicle to mature, which then in turn secretes estrogen. Unlike most mammals, the mare does not have an increase of LH right before ovulation.

FSH (Follicle-stimulating hormone): secreted by the pituitary, causes the ovarian follicle to develop. Levels of FSH rise slightly at the end of estrus, but have their highest peak about 10 days before the next ovulation. FSH is inhibited by inhibin, at the same time LH and estrogen levels rise, which prevents immature follicles from continuing their growth. Mares may however have multiple FSH waves during a single estrous cycle, and diestrus follicles resulting from a diestrus FSH wave are not uncommon, particularly in the height of the natural breeding season.

Estrogen: secreted by the developing follicle, it causes the pituitary gland to secrete more LH (therefore, these 2 hormones are in a positive feedback loop). Additionally, it causes behavioural changes in the mare, making her more receptive toward the stallion, and causes physical changes in the cervix, uterus, and vagina to prepare the mare for conception. Estrogen peaks 1–2 days before ovulation, and decreases within 2 days following ovulation.

Inhibin: secreted by the developed follicle right before ovulation, "turns off" FSH, which is no longer needed now that the follicle is larger.

Progesterone: prevents conception and decreases sexual receptibility of the mare to the stallion. Progesterone is therefore lowest during the estrus phase, and increases during diestrus. It decreases 12–15 days after ovulation, when the corpus luteum begins to decrease in size.

Prostaglandin: secreted by the endrometrium 13–15 days following ovulation, causes luteolysis and prevents the corpus luteum from secreting progesterone

CG-equine chorionic gonadotropin-(also called PMSG (pregnant mare serum gonadotropin): chorionic gonadotropins secreted if the mare conceives. First secreted by the endometrial cups around the 36th day of gestation, peaking around day 60, and decreasing after about 120 days of gestation. Also help to stimulate the growth of the fetal gonads.

Breeding and Gestation

While horses in the wild mate and foal in mid to late spring, in the case of horses domestically bred for competitive purposes, especially horse racing and various futurities, it is desirable that they be born as close to January 1 in the northern hemisphere or August 1 in the southern hemisphere as possible, so as to be at an advantage in size and maturity when competing against other horses in the same age group. When an early foal is desired, barn managers will put the mare "under lights" by keeping the barn lights on in the winter to simulate a longer day, thus bringing the mare into estrus sooner than she would in nature. Mares signal estrus and ovulation by urination in the presence of a stallion, raising the tail and revealing the vulva. A stallion, approaching with a high head, will usually nicker, nip and nudge the mare, as well as sniff her urine to determine her readiness for mating.

Once fertilized, the oocyte (egg) remains in the oviduct for approximately 5.5 more days, and then descends into the uterus. The initial single cell combination is already dividing and by the time of entry into the uterus, the egg might have already reached the blastocyst stage. The gestation period lasts for about eleven months, or about 340 days (normal average range 320–370 days). During the early days of pregnancy, the conceptus is mobile, moving about in the uterus until about day 16 when "fixation" occurs. Shortly after fixation, the embryo proper (so called up to about 35 days) will become visible on trans-rectal ultrasound (about day 21) and a heartbeat should be visible by about day 23. After the formation of the endometrial cups and early placentation is initiated (35–40 days of gestation) the terminology changes, and the embryo is referred to as a fetus. True implantation-invasion into the endometrium of any sort-does not occur until about day 35 of pregnancy with the formation of the endometrial cups, and true placentation (formation of the placenta) is not initiated until about day 40-45 and not completed until about 140 days of pregnancy. The fetus gender can be determined by day 70 of the gestation using ultrasound. Halfway through gestation the fetus is the size of between a rabbit and a beagle. The most dramatic fetal development occurs in the last 3 months of pregnancy when 60% of fetal growth occurs.

Colts are carried on average about 4 days longer than fillies.

Care of the Pregnant Mare

Domestic mares receive specific care and nutrition to ensure that they and their foals are healthy. Mares are given vaccinations against diseases such as the Rhinopneumonitis (EHV-1) virus (which can cause abortions) as well as vaccines for other conditions that may occur in a given region of the world. Pre-foaling vaccines are recommended 4–6 weeks prior to foaling to maximize the immunoglobulin content of the colostrum in the first milk. Deworming the mare a few weeks prior to foaling is also important, as the mare is the primary source of parasites for the foal.

Mares can be used for riding or driving during most of their pregnancy, and it's healthy for them to have exercise. But only moderate exercise, especially when they become heavy in foal. Exercise in excessively high temperatures has been suggested as being detrimental to pregnancy maintenance during the embryonic period-it should however be noted that ambient temperatures encountered during the research were in the region of 100 degrees F and the same results may not be encountered in regions with lower ambient temperatures.

During the last 3–4 months of gestation, rapid growth of the fetus increases the pregnant mare's nutritional requirements. Energy requirements during these last few months, and during the first few months of lactation are similar to those of a horse in full training. Trace minerals such as Copper are extremely important, particularly during the tenth month of pregnancy, for proper skeletal formation. Many feeds designed for pregnant and lactating mares provide the careful balance required of increased protein, increased calories through extra fat as well as vitamins and minerals. During the first several months of pregnancy, the nutritional requirements do not increase significantly since the rate of growth of the fetus is very slow. However, during this time, the mare should be provided supplemental vitamins and minerals, particularly if forage quality is questionable. Overfeeding the pregnant mare, particularly during early gestation, should be avoided, as excess weight may contribute to difficulties foaling or fetal/foal related problems.

Foaling

Mares due to foal are usually separated from other horses, both for the benefit of the mare and the safety of the soon-to-be-delivered foal. In addition, separation allows the mare to be monitored more closely by humans for any problems that may occur while giving birth. In the northern hemisphere a special foaling stall that is large and clutter free is frequently used, particularly by major breeding farms. Originally, this was due in part to a need for protection from the harsh winter climate present when mares foal early in the year, but even in moderate climates, such as Florida, foaling stalls are still common because they allow closer monitoring of mares. Smaller breeders often use a small pen with a large shed for foaling, or they may remove a wall between two box stalls in a small barn to make a large stall. In the milder climates seen in much of the southern hemisphere, most mares foal outside, often in a paddock built specifically for foaling, especially on the larger stud farms. Many stud farms worldwide employ technology to alert human managers when the mare is about to foal, including webcams, closed-circuit television, or assorted types of devices that alert a handler via a remote alarm when a mare lies down in a position to foal.

On the other hand, some breeders, particularly those in remote areas or with extremely large numbers of horses, may allow mares to foal out in a field amongst a herd, but may also see higher rates of foal and mare mortality in doing so. Most mares foal at night or early in the morning, and prefer to give birth alone when possible.

Labour is rapid, often no more than 30 minutes, and from the time the feet of the foal appear to full delivery is often only about 15 to 20 minutes. Once the foal is born, the mare will lick the newborn foal to clean it and help blood circulation. In a very short time, the foal will attempt to stand and get milk from its mother. A foal should stand and nurse within the first hour of life.

To create a bond with her foal, the mare licks and nuzzles the foal, enabling her to distinguish hers from others. Some mares are aggressive when protecting their foals, and may attack other horses or unfamiliar humans that come near their newborns.

After birth, a foal's navel is dipped in antiseptic to prevent infection, it is sometimes given an enema to help clear the meconium from its digestive tract, and the newborn is monitored to ensure that it stands and nurses without difficulty. While most horse births happen without complications, many owners have first aid supplies prepared and a veterinarian on call in case of a birthing emergency. People who supervise foaling should also watch the mare to be sure that she passes the placenta in a timely fashion, and that it is complete with no fragments remaining in the uterus, where retained fetal membranes could cause a serious inflammatory condition (endometritis) and/or infection. If the placenta is not removed from the stall after it is passed, a mare will often eat it, an instinct from the wild, where blood would attract predators.

Foal Care

A foal with its mother, or damFoals develop rapidly, and within a few hours a wild foal can travel with the herd. In domestic breeding, the foal and dam are usually separated from the herd for a while, but within a few weeks are typically pastured with the other horses. A foal will begin to eat hay, grass and grain alongside the mare at about 4 weeks old; by 10–12 weeks the foal requires more nutrition than the mare's milk can supply. Foals are typically weaned at 4–8 months of age, although in the wild a foal may nurse for a year.

How Breeds Bevelop

Beyond the appearance and conformation of a specific type of horse, breeders aspire to improve physical performance abilities. This concept, known as matching "form to function," has led to the development of not only different breeds, but also families or bloodlines within breeds that are specialists for excelling at specific tasks.

For example, the Arabian horse of the desert naturally developed speed and endurance to travel long distances and survive in a harsh environment, and domestication by humans added a trainable disposition to the animal's natural abilities. In the meantime, in northern Europe, the locally adapted heavy horse with a thick, warm coat was domesticated and put to work as a farm animal that could pull a plow or wagon. This animal was later adapted through selective breeding to create a strong but ridable animal suitable for the heavily-armored knight in warfare.

Then, centuries later, when people in Europe wanted faster horses than could be produced from local horses through simple selective breeding, they imported Arabians and other oriental horses to breed as an outcross to the heavier, local animals. This led to the development of breeds such as the Thoroughbred, a horse taller than the Arabian and faster over the distances of a few miles required of a European race horse or light cavalry horse. Another cross between oriental and European horses produced the Andalusian, a horse developed in Spain that was powerfully built, but extremely nimble and capable of the quick bursts of speed over short distances necessary for certain types of combat as well as for tasks such as bullfighting.

Later, the people who settled the Americas needed a hardy horse that was capable of working with cattle. Thus, Arabians and Thoroughbreds were crossed on Spanish horses, both domesticated animals descended from those brought over by the Conquistadors, and feral horses such as the Mustangs, descended from the Spanish horse, but adapted by natural selection to the ecology and climate of the west. These crosses ultimately produced new breeds such as the American quarter horse and the Criollo of Argentina.

In modern times, these breeds themselves have since been selectively bred to further specialize at certain tasks. One example of this is the American quarter horse. Once a general-purpose working ranch horse, different bloodlines now specialize in different events. For example, larger, heavier animals with a very steady attitude are bred to give competitors an advantage in events such as team roping, where a horse has to start and stop quickly, but also must calmly hold a full-grown steer at the end of a rope. On the other hand, for an event known as cutting, where the horse must separate a cow from a herd and prevent it from rejoining the group, the best horses are smaller, quick, alert, athletic and highly trainable. They must learn quickly, have conformation that allows quick stops and fast, low turns, and the best competitors have a certain amount of independent mental ability to anticipate and counter the movement of a cow, popularly known as "cow sense."

Another example is the Thoroughbred. While most representatives of this breed are bred for horse racing, there are also specialized bloodlines suitable as show hunters or show jumpers. The hunter must have a tall, smooth build that allows it to trot and canter smoothly and efficiently. Instead of speed, value is placed on appearance and upon giving the equestrian a comfortable ride, with natural jumping ability that shows bascule and good form.

A show jumper, however, is bred less for overall form and more for power over tall fences, along with speed, scope, and agility. This favors a horse with a good galloping stride, powerful hindquarters that can change speed or direction easily, plus a good shoulder angle and length of neck. A jumper has a more powerful build than either the hunter or the racehorse.

History of Horse Breeding

The history of horse breeding goes back millennia. Though the precise date is in dispute, humans could have domesticated the horse as far back as approximately 4500 BCE. However, evidence of planned breeding has a more blurry history.

One of the earliest people known to document the breedings of their horses were the Bedouin of the Middle East, the breeders of the Arabian horse. While it is difficult to determine how far back the Bedouin passed on pedigree information via an oral tradition, there were written pedigrees of Arabian horses by A.D. 1330. The Akhal-Teke of West-Central Asia is another breed with roots in ancient times that was also bred specifically for war and racing. The nomads of the Mongolian steppes bred horses for several thousand years as well.

The types of horse bred varied with culture and with the times. The uses to which a horse was put also determined its qualities, including smooth amblers for riding, fast horses for carrying messengers, heavy horses for plowing and pulling heavy wagons, ponies for hauling cars of ore from mines, packhorses, carriage horses and many others.

Medieval Europe bred large horses specifically for war, called destriers. These horses were the ancestors of the great heavy horses of today, and their size was preferred not simply because of the weight of the armor, but also because a large horse provided more power for the knight's lance. Weighing almost twice as much as a normal riding horse, the destrier was a powerful weapon in battle.

On the other hand, during this same time, lighter horses were bred in northern Africa and the Middle East by Muslim warriors, who preferred a faster, more agile horse. The lighter horse suited the raids and battles of the Bedouins, allowing them to outmaneuver rather than overpower the enemy. When Muslim warriors and European knights collided in warfare, the heavy knights were frequently outmaneuvered. The Europeans, however, soon made up for the lack of speed of their native breeds by incorporating genetic traits from captured oriental horses such as the Arabian, Barb to their stables. This cross-breeding led both to a nimbler war horse, such as today's Percheron, but also to created a type of horse known as a Courser, a predecessor to the Thoroughbred, which was used as a message horse.

During the Renaissance, horses were bred not only for war, but for haute ecole riding, derived from the most athletic movements required of a war horse, and popular among the elite nobility of the time. Breeds such as the Lipizzan were developed from Spanish-bred horses for this purpose, and also became the preferred mounts of cavalry officers, who were derived mostly from the ranks of the nobility. It was during this time that gunpowder was developed, and so the light cavalry horse, a faster and quicker war horse, was bred for a "shoot and run" tactic rather than the close hand-to-hand fighting seen in the Middle Ages.

After Charles II retook the British throne in 1660, horse racing, which had been banned by Cromwell, was revived. The Thoroughbred was developed 40 years later, bred to be the ultimate racehorse, through the lines of 3 foundation Arabian stallions.

In the 18th century, James Burnett, Lord Monboddo noted the importance of selecting appropriate parentage to achieve desired outcomes of successive generations. Monboddo worked more broadly in the abstract thought of species relationships and evolution of species. The Thoroughbred breeding hub in Lexington, Kentucky was developed in the late 18th century, and became a mainstay in American racehorse breeding. The 17th and 18th centuries saw more of a need for fine carriage horses in Europe, bringing in the dawn of the warmblood.

The warmblood breeds have been exceptionally good at adapting to changing times, and from their carriage horse beginnings they easily transitioned during the 20th century into a sport horse type. Today's warmblood breeds, although still used for competitive driving, are more often seen competing in the show jumping or dressage arenas.

The Thoroughbred continues to dominate the horseracing world, although its lines have been more recently used to improve warmblood breeds and to develop sport horses.

The predecessor of the American Quarter Horse was developed in the 18th century, mainly for quarter racing (racing ¼ of a mile). The breed was later adapted for work in the west, and "cow sense" was particularly bred for as their use for herding cattle increased. However, because there was also a need for animals suitable for sprint racing, the modern Quarter Horse has two distinct types: the sleeker racing type and the stock horse type. The racing type most resembles the finer-boned ancestors of the first racing Quarter Horses, and the type is still used for ¼-mile races. The stock horse type, used in western events, is bred for a shorter stride, docile temperament, and cow sense.

The need for horses for heavy draft and carriage work continued until the industrial revolution and the advent of the automobile and the tractor. After this time, draft and carriage horse numbers dropped significantly, though light riding horses remained popular for recreational pursuits. Draft horses today are used on a few small farms, but today are seen mainly for pulling and plowing competitions rather than farm work. Heavy harness horses are now used as an outcross with lighter breeds, such as the Thoroughbred, to produce the modern warmblood breeds popular in Olympic and sport horse disciplines.

Deciding to Breed a Horse

Breeding a horse is an endeavour where the owner, particularly of the mare, will usually need to invest considerable time and money. For this reason, a horse owner needs to consider several factors, including:

Does the proposed breeding animal have valuable genetic qualities to pass on?

Is the proposed breeding animal in good physical health, fertile, and able to withstand the vigors of reproduction?

For what purpose will the foal be used?

Is there a market for the foal in the event that the owner does not wish to keep the foal for its entire life?

What is the anticipated economic benefit, if any, to the owner of the ensuing foal?

What is the anticipated economic benefit, if any, to the owner(s) of the sire and dam or the foal?

Does the owner of the mare have the expertise to properly manage the mare through gestation and parturition?

Does the owner of the potential foal have the expertise to properly manage and train a young animal once it is born?

There are value judgements involved in considering whether an animal is suitable breeding stock, hotly debated by breeders. Additional personal beliefs may come into play when considering a suitable level of care for the mare and ensuing foal, the potential market or use for the foal, and other tangible and intangible benefits to the owner.

If the breeding endeavour is intended to make a profit, there are additional market factors to consider, which may vary considerably from year to year, from breed to breed, and by region of the world. In many cases, the low end of the market is saturated with horses, and the law of supply and demand thus allows little or no profit to be made from breeding unregistered animals or animals of poor quality, even if registered.

The minimum cost of breeding for a mare owner includes the stud fee, and the cost of proper nutrition, management and veterinary care of the mare throughout gestation, parturition, and care of both mare and foal up to the time of weaning. Veterinary expenses may be higher if specialized reproductive technologies are used or health complications occur.

Making a profit in horse breeding is often difficult. While some owners of only a few horses may keep a foal for purely personal enjoyment, many individuals breed horses in hopes of making some money in the process.

A general rule of thumb is that a foal intended for sale should be worth three times the cost of the stud fee if it were sold at the moment of birth. From birth forward, the costs of care and training are added to the value of the foal, with a sale price going up accordingly. If the foal wins awards in some form of competition, that may also enhance the price. On the other hand, without careful thought, foals bred without a potential market for them may wind up being sold at a loss, and in a worst-case scenario, sold for "salvage" value—a euphemism for sale to slaughter as horsemeat.

Therefore, a mare owner must consider their reasons for breeding, asking hard questions of themselves as to whether their motivations are based on either emotion or profit and how realistic those motivations may be.

Choosing Breeding Stock

A stallion with a proven competition record is one criterion for being a suitable sire.

The stallion should be chosen to complement the mare, with the goal of producing a foal that has the best qualities of both animals, yet avoids having the weaker qualities of either parent. Generally, the stallion should have proven himself in the discipline or sport the mare owner wishes for the "career" of the ensuing foal. Mares should also have a competition record showing that they also have suitable traits, though this does not happen as often.

Some breeders consider the quality of the sire to be more important than the quality of the dam. However, other breeders maintain that the mare is the most important parent. Because stallions can produce far more offspring than mares, a single stallion can have a greater overall impact on a breed.

However, the mare may have a greater influence on an individual foal because its physical characteristics influence the developing foal in the womb and the foal also learns habits from its dam when young. Foals may also learn the "language of intimidation and submission" from their dam, and this imprinting may affect the foal's status and rank within the herd. Many times, a mature horse will achieve status in a herd similar to that of its dam; the offspring of dominant mares become dominant themselves.

A purebred horse is usually worth more than a horse of mixed breeding, though this matters more in some disciplines than others. The breed of the horse is sometimes secondary when breeding for a sport horse, but some disciplines may prefer a certain breed or a specific phenotype of horse. Sometimes, purebred bloodlines are an absolute requirement: For example most Racehorses in the world must be recorded with a breed registry in order to race.

Bloodlines are often considered, as some bloodlines are known to cross well with others. If the parents have not yet proven themselves by competition or by producing quality offspring, the bloodlines of the horse are often a good indicator of quality and possible strengths and weaknesses. Some bloodlines are known not only for their athletic ability, but could also carry a conformational or genetic defect, poor temperament, or for a medical problem. Some bloodlines are also fashionable or otherwise marketable, which is an important consideration should the mare owner wish to sell the foal.

Horse breeders also consider conformation, size and temperament. All of these traits are heritable, and will determine if the foal will be a success in its chosen discipline. The offspring, or "get", of a stallion are often excellent indicators of his ability to pass on his characteristics, and the particular traits he actually passes on. Some stallions are fantastic performers but never produce offspring of comparable quality. Others sire fillies of great abilities but not colts. At times, a horse of mediocre ability sires foals of outstanding quality.

Mare owners also look into the question of if the stallion is fertile and has successfully "settled" (i.e. impregnated) mares. A stallion may not be able to breed naturally, or old age may decrease his performance. Mare care boarding fees and semen collection fees can be a major cost.

Costs Related to Breeding

Breeding a horse can be an expensive endeavour, whether breeding a backyard competition horse or the next Olympic medalist. Costs may include:

The stud and booking fee:

- Fees for collecting, handling, and transporting semen (if AI is used and semen is shipped)
- Mare exams: to determine if she is healthy enough to breed, to determine when she ovulates, and (if AI is used) to inseminate her
- Mare transport, care, and board if the mare is bred live cover at the stallion's residence
- Veterinary bills to keep the pregnant mare healthy while in foal
- Possible veterinary bills during pregnancy or foaling should something go wrong

- Veterinary bills for the foal for its first exam a few days following foaling.

Stud fees are determined by the quality of the stallion, his performance record, the performance record of his get (offspring), as well as the sport and general market that the animal is standing for.

The highest stud fees are generally for racing Thoroughbreds, which may charge from two to three thousand dollars for a breeding to a new or unproven stallion, to several hundred thousand dollars for a breeding to a proven producer of stakes winners. Stallions in other disciplines often have stud fees that begin in the range of $1000 to $3000, with top contenders who produce champions in certain disciplines able to command as much as $20,000 for one breeding. The lowest stud fees to breed to a grade horse or an animal of low-quality pedigree may only be $100–$200, but there are trade-offs: the horse will probably be unproven, and likely to produce lower-quality offspring than a horse with a stud fee that is in the typical range for quality breeding stock.

As a stallion's career, either performance or breeding, improves, his stud fee tends to increase in proportion. If one or two offspring are especially successful, winning several stakes races or an Olympic medal, the stud fee will generally greatly increase. Younger, unproven stallions will generally have a lower stud fee earlier on in their careers.

To help decrease the risk of financial loss should the mare die or abort the foal while pregnant, many studs have a live foal guarantee (LFG)-also known as "no foal, free return" or "NFFR"-allowing the owner to have a free breeding to their stallion the next year. However, this is not offered for every breeding.

Covering the Mare

An artificial vagina, used to collect semen. There are two general ways to "cover" or breed the mare: Live cover: the mare is brought to the stallion's residence and is covered "live" in the breeding shed. She may also be turned out in a pasture with the stallion for several days to breed naturally ('pasture bred'). The former situation is often preferred, as it provides a more controlled environment, allowing the breeder to ensure that the mare was covered, and places the handlers in a position to remove the horses from one another should one attempt to kick or bite the other.

Artificial Insemination (AI): the mare is inseminated by a veterinarian or an equine reproduction manager, using either fresh, cooled or frozen semen.

After the mare is bred or artificially inseminated, she is checked using ultrasound 14–16 days later to see if she "took", and is pregnant. A second check is usually performed at 28 days. If the mare is not pregnant, she may be bred again during her next cycle.

It is considered safe to breed a mare to a stallion of much larger size. Because of the mare's type of placenta and its attachment and blood supply, the foal will be limited in its growth within the uterus to the size of the mare's uterus, but will grow to its genetic potential after it is born. Test breedings have been done with draft horse stallions bred to small mares with no increase in the number of difficult births.

Live Cover

When breeding live cover, the mare is usually boarded at the stud. She is "teased" several times with a stallion that will not breed to her, usually with the stallion being presented to the mare over a barrier. Her reaction to the teaser, whether hostile or passive, is noted. A mare that is in heat will generally tolerate a teaser (although this is not always the case), and may present herself to him, holding her tail to the side. A veterinarian may also determine if the mare is ready to be bred, by ultrasound or palpating daily to determine if ovulation has occurred.

When it has been determined that the mare is ready, both the mare and intended stud will be cleaned. The mare will then be presented to the stallion, usually with one handler controlling the mare and one or more handlers in charge of the stallion. Multiple handlers are preferred, as the mare and stallion can be easily separated should there be any trouble.

The Jockey Club, the organization that oversees the Thoroughbred industry in the United States, requires all registered foals to be bred through live cover. Artificial insemination, listed below, is not permitted. Similar rules apply in other countries.

By contrast, the U.S. standardbred industry allows registered foals to be bred by live cover, or by artificial insemination (AI) with fresh or frozen (not dried) semen. No other artificial fertility treatment is allowed. In addition, foals bred via AI of frozen semen may only be registered if the stallion's sperm was collected during his lifetime, and used no later than the calendar year of his death or castration.

Artificial Insemination

Artificial insemination (AI) has several advantages over live cover, and has a very similar conception rate:

The mare and stallion never have to come in contact with each other, which therefore reduces breeding accidents, such as the mare kicking the stallion.

AI opens up the world to international breeding, as semen may be shipped across continents to mares that would otherwise be unable to breed to a particular stallion.

A mare also does not have to travel to the stallion, so the process is less stressful on her, and if she already has a foal, the foal does not have to travel.

AI allows more mares to be bred from one stallion, as the ejaculate may be split between mares.

AI reduces the chance of spreading sexually transmitted diseases between mare and stallion.

AI allows mares or stallions with health issues, such as sore hocks which may prevent a stallion from mounting, to continue to breed.

Frozen semen may be stored and used to breed mares even after the stallion is dead, allowing his lines to continue. However, the semen of some stallions does not freeze well. Some breed registries may not permit the registration of foals resulting from the use of frozen semen after the stallion's death, although other large registries accept such usage and provide registrations. The overall trend is toward permitting use of frozen semen after the death of the stallion.

A stallion is usually trained to mount a phantom (or dummy) mare, although a live mare may be used, and he is most commonly collected using an artificial vagina (AV) which is heated to simulate the vagina of the mare. The AV has a filter and collection area at one end to capture the semen, which can then be processed in a lab. The semen may be chilled or frozen and shipped to the mare owner or used to breed mares "on-farm". When the mare is in heat, the person inseminating introduces the semen directly into her uterus using a syringe and pipette.

Advanced Reproductive Techniques

Often an owner does not want to take a valuable competition mare out of training to carry a foal. This presents a problem, as the mare will usually be quite old by the time she is retired from her competitive career, at which time it is more difficult to impregnate her. Other times, a mare may have physical problems that prevent or discourage breeding. However, there are now several options for breeding these mares. These options also allow a mare to produce multiple foals each breeding season, instead of the usual one. Therefore, mares may have an even greater value for breeding.

Embryo Transfer: This relatively new method involves flushing out the mare's fertilized embryo a few days following insemination, and transferring to a surrogate mare, which has been synchronized to be in the same phase of the estrous cycle as the donor mare.

Gamete Intrafallopian Transfer (GIFT): The mare's ovum and the stallion's sperm are deposited in the oviduct of a surrogate dam. This technique is very useful for subfertile stallions, as fewer sperm are needed, so a stallion with a low sperm count can still successfully breed.

Egg Transfer: An oocyte is removed from the mare's follicle and transferred into the oviduct of the recipient mare, who is then bred. This is best for mares with physical problems, such as an obstructed oviduct, that prevent breeding.

Inbreeding

Inbreeding is the reproduction from the mating of two genetically related parents. Inbreeding results in increased homozygosity, which can increase the chances of offspring being affected by recessive or deleterious traits. This generally leads to a decreased fitness of a population, which is called inbreeding depression. Deleterious alleles causing inbreeding depression can subsequently be removed through culling, which is also known as genetic purging.

Livestock breeders often practice controlled breeding to eliminate undesirable characteristics within a population, which is also coupled with culling of what is considered unfit offspring, especially when trying to establish a new and desirable trait in the stock.

In plant breeding, inbred lines are used as stocks for the creation of hybrid lines to make use of the effects of heterosis. Inbreeding in plants also occurs naturally in the form of self-pollination.

Results

Inbreeding may result in a far higher phenotypic expression of deleterious recessive genes within a population than would normally be expected. As a result, first-generation inbred individuals are more likely to show physical and health defects, including:

- Reduced fertility both in litter size and sperm viability
- Increased genetic disorders
- Fluctuating facial asymmetry
- Lower birth rate
- Higher infant mortality
- Slower growth rate
- Smaller adult size
- Loss of immune system function.

Natural selection works to remove individuals with the above types of traits from the gene pool. Therefore, many more individuals in the first generation of inbreeding will never live to reproduce. Over time, with isolation such as a population bottleneck caused by purposeful (assortative) breeding or natural environmental stresses, the deleterious inherited traits are culled.

Island species are often very inbred, as their isolation from the larger group on a mainland allows for natural selection to work upon their population. This type of isolation may result in the formation of race or even speciation, as the inbreeding first removes many deleterious genes, and allows expression of genes that allow a population to adapt to an ecosystem. As the adaptation becomes more pronounced the new species or race radiates from its entrance into the new space, or dies out if it cannot adapt and, most importantly, reproduce.

The reduced genetic diversity that results from inbreeding may mean a species may not be able to adapt to changes in environmental conditions. Each individual will have similar immune systems, as immune systems are genetically based. Where a species becomes endangered, the population may fall below a minimum whereby the forced interbreeding between the remaining animals will result in extinction.

Natural breedings include inbreeding by necessity, and most animals only migrate when necessary. In many cases, the closest

available mate is a mother, sister, grandmother, father, grandfather...
In all cases the environment presents stresses to remove those
individuals who cannot survive because of illness from the population.

There was an assumption that wild populations do not inbreed;
this is not what is observed in some cases in the wild. However, in
species such as horses, animals in wild or feral conditions often drive
off the young of both genders, thought to be a mechanism by which
the species instinctively avoids some of the genetic consequences of
inbreeding. In general, many mammal species including humanity's
closest primate relatives avoid close inbreeding possibly due to the
deleterious effects.

Examples

The cheetah was once reduced by disease, habitat restriction,
overhunting of prey, competition from other predators (primarily
lions, competition from human land use, etc.) to a very small number
of individuals. All cheetahs now come from this very small gene pool.
Should a virus appear that none of the cheetahs have resistance to,
extinction is always a possibility. Currently, the threatening virus is
feline infectious peritonitis, which has a disease rate in domestic cats
from 1%–5%; in the cheetah population it is ranging between 50% to
60%. The cheetah is also known, in spite of its small gene pool, for
few genetic illnesses.

In the South American sea lion, there was concern that recent
population crashes would reduce genetic diversity. Historical analysis
indicated that a population expansion from just two matrilineal lines
were responsible for most individuals within the population. Even so,
the diversity within the lines allowed for great variation in the gene
pool that may help to protect the South American sea lion from
extinction.

In lions, prides are often followed by related males in bachelor
groups. When the dominant male is killed or driven off by one of these
bachelors, a father may be replaced with his son. There is no mechanism
for preventing inbreeding or to ensure outcrossing. In the prides, most
lionesses are related to one another. If there is more than one dominant
male, the group of alpha males are usually related. Two lines are then
being "line bred". Also, in some populations such as the Crater lions,
it is known that a population bottleneck has occurred. Researchers
found far greater genetic heterozygosity than expected. In fact,
predators are known for low genetic variance, along with most of the

top portion of the tropic levels of an ecosystem. Additionally, the alpha males of two neighbouring prides can potentially be from the same litter; one brother may come to acquire leadership over another's pride, and subsequently mate with his 'nieces' or cousins. However, killing another male's cubs, upon the takeover, allows for the new selected gene complement of the incoming alpha male to prevail over the previous male. There are genetic assays being scheduled for lions to determine their genetic diversity. The preliminary studies show results inconsistent with the outcrossing paradigm based on individual environments of the studied groups.

Calculation

The inbreeding is computed as a percentage of chances for two alleles to be identical by descent. This percentage is called "inbreeding coefficient". There are several methods to compute this percentage, the two main ways are the path method and the tabular method.

Typical inbreeding percentages are as follows, assuming no previous inbreeding between any parents:

- Father/daughter, mother/son or brother/sister '! 25%
- Grandfather/granddaughter or grandmother/grandson '! 12.5%
- Half-brother/half-sister '! 12.5%
- Uncle/niece or aunt/nephew '! 12.5%
- Great-grandfather/great-granddaughter or great-grandmother/ great-grandson '! 6.25%
- Half-uncle/niece or half-aunt/nephew '! 6.25%
- First cousins '! 6.25%
- First cousins once removed or half-first cousins '! 3.125%
- Second cousins or first cousins twice removed '! 1.5625%
- Second cousins once removed or half-second cousins '! 0.78125%.

An inbreeding calculation may be used to determine the general genetic distance among relatives by multiplying by two, because any progeny would have a 1 in 2 risk of actually inheriting the identical alleles from both parents.

For instance, the parent/child or sibling/sibling relationships have 50% identical genetics.

Note: For siblings, the degree of genetic relationship is not an automatic 50% as it is with parents and their children, but a range from 100% at one extreme, as in the case of identical twins (who obviously cannot mate as they are the same sex), to an exceedingly unlikely 0%. In other words, siblings share an average of 50% of their genes, but unlike the 50% ratio between parents and children, the actual ratio between siblings in any given case can vary

An intensive form of inbreeding where an individual S is mated to his daughter D1, granddaughter D2 and so on, in order to maximise the percentage of S's genes in the offspring. D3 would have 87.5% of his genes, while D4 would have 93.75%.

Breeding in domestic animals is assortative breeding primarily. Without the sorting of individuals by trait, a breed could not be established, nor could poor genetic material be removed. Homozygosity is the case where similar or identical alleles combine to express a trait that is not otherwise expressed (recessiveness). Inbreeding, through homozygosity, exposes recessive alleles. Inbreeding is used to reveal deleterious recessive alleles, which can then be eliminated through assortative breeding or through culling. Inbreeding is used by breeders of domestic animals to fix desirable genetic traits within a population or to attempt to remove deleterious traits by allowing them to manifest phenotypically from the genotypes. Inbreeding is defined as the use of close relations for breeding such as mother to son, father to daughter, brother to sister. Breeders must cull unfit breeding suppressed individuals and/or individuals who demonstrate either homozygosity or heterozygosity for genetic based diseases. The issue of casual breeders who inbreed irresponsibly is discussed in the following quotation on cattle: Meanwhile, milk production per cow per lactation increased from 17,444 lbs to 25,013 lbs from 1978 to 1998 for the Holstein breed. Mean breeding values for milk of Holstein cows increased by 4,829 lbs during this period. High producing cows are increasingly difficult to breed and are subject to higher health costs than cows of lower genetic merit for production (Cassell, 2001). Intensive selection for higher yield has increased relationships among animals within breed and increased the rate of casual inbreeding. Many of the traits that affect profitability in crosses of modern dairy breeds have not been studied in designed experiments. Indeed, all crossbreeding research involving North American breeds and strains is very dated (McAllister, 2001) if it exists at all.

Linebreeding is a form of inbreeding. There is no clear distinction between the two terms, but linebreeding may encompass crosses between individuals and their descendants or two cousins. This method can be used to increase a particular animal's contribution to the population. While linebreeding is less likely to cause problems in the first generation than does inbreeding, over time, linebreeding can reduce the genetic diversity of a population and cause problems related to a too-small genepool that may include an increased prevalence of genetic disorders and inbreeding depression.

Outcrossing is where two unrelated individuals have been crossed to produce progeny. In outcrossing, unless there is verifiable genetic information, one may find that all individuals are distantly related to an ancient progenitor. If the trait carries throughout a population, all individuals can have this trait. This is called the founder's effect. In the well established breeds, that are commonly bred, a large gene pool is present. For example, in 2004, over 18,000 Persian cats were registered. A possibility exists for a complete outcross, if no barriers exist between the individuals to breed. However it is not always the case, and a form of distant linebreeding occurs. Again it is up to the assortative breeder to know what sort of traits both positive and negative exist within the diversity of one breeding. This diversity of genetic expression, within even close relatives, increases the variability and diversity of viable stock.

The two dog sites above also point out that in the registered dog population, the onset of large numbers of casual breeders has corresponded with an increase in the number of genetic illnesses of dogs by not understanding how, why and which traits are inherited. The dog sites indicate that the largest percentage of dog breeders in the US are casual breeders. Therefore the investment in a papered animal, with an expected short term profit, motivates some to ignore the practice of culling. Casual breeders in companion animals often ignore breeding restrictions within their contracts with source companion animal breeders. The casual breeders breed the very culls that a genetics based breeder has released as a pet. The casual breeder was also cited in the quotes above on cattle raising.

Laboratory Animals

Systematic inbreeding and maintenance of inbred strains of laboratory mice and rats is of great importance for biomedical research. The inbreeding guarantees a consistent and uniform animal model for experimental purposes and enables genetic studies in congenic and knock-out animals. The use of inbred strains is also important for genetic studies in animal models, for example to distinguish genetic from environmental effects.

Genetic Disorders

The offspring of consanguineous relationships are at greater risk of certain genetic disorders. Autosomal recessive disorders occur in individuals who are homozygous for a particular recessive gene mutation. This means that they carry two copies of the same gene (allele). Except in certain rare circumstances (new mutations or uniparental disomy) both parents of an individual with such a disorder will be carriers of the gene.

Such carriers are not affected and will not display any signs that they are carriers, and so may be unaware that they carry the mutated gene. As relatives share a proportion of their genes, it is much more likely that related parents will be carriers of the same autosomal recessive gene, and therefore their children are at a higher risk of an autosomal recessive disorder. The extent to which the risk increases depends on the degree of genetic relationship between the parents; so the risk is greater in mating relationships where the parents are close relatives, but for relationships between more distant relatives, such as second cousins, the risk is lower (although still greater than the general population). A 1994 study found the progeny of first cousins, in Pakistan, indicate morbidity levels to be some 1% to 4% higher than in the offspring of unrelated couples. They went on to report, however, that this number was significantly inflated by sociodemographic variables.

Prohibitions to Inbreeding

The taboo of incest has been discussed by many social scientists. As anthropologists attest, this taboo exists in most cultures. As inbreeding within the first generation often produces expression of recessive traits, the prohibition has been discussed as a possible functional response to the requirement of culling those born deformed, or with undesirable traits. Some biologists like Charles Davenport advocated traditional forms of assortative breeding, *i.e.*, eugenics, to form better "human stock".

Some Hindus follow the Gotra system, which prescribes prohibition of marriages among relatives based on a name attached to paternal relatives, to prevent inbreeding. Direct inbreeding is also prohibited in Islam, as described in the Quran (chapter 4, verse 23).

Royalty and Nobility

The family relationships of royalty are usually very well known, leading observers to view royalty as highly *inbred*, but they are often comparable to many ethnic groups where the relationships are not publicized as well. Royal intermarriage was often practised to protect property, wealth, and position.

In ancient Egypt, royal women carried the bloodlines and so it was advantageous for a pharaoh to marry his sister or half-sister; in such cases a special combination between endogamy and polygamy is found. Normally the old ruler's eldest son and daughter (who could be either siblings or half-siblings) became the new rulers. All rulers of the Ptolemaic dynasty from Ptolemy II were married to their brothers and sisters, so as to keep the Ptolemaic blood "pure" and to strengthen the line of succession. Cleopatra VII (also called Cleopatra VI) and Ptolemy XIII, who married and became co-rulers of ancient Egypt following their father's death, are the most widely known example.

Among European monarchies Jean V of Armagnac formed a rare brother-sister relationship. Also other royal houses, such as the Wittelsbachs had marriages among aunts, uncles, nieces, and nephews. The British royal family had several marriages as close as the first cousin, but none closer.

The most famous example of a genetic disorder aggravated by royal family intermarriage was the House of Habsburg, which inmarried particularly often. Famous in this case is the *Habsburger (Unter) Lippe* (Habsburg jaw/Habsburg lip/"Austrian lip") (mandibular prognathism), typical for many Habsburg relatives over a period of six centuries. The condition progressed through the generations to the point that the last of the Spanish Habsburgs, Charles II of Spain, could not properly chew his food.

Besides the jaw deformity, Charles II also had a huge number of other genetic physical, intellectual, sexual, and emotional problems. It is speculated that the simultaneous occurrence in Charles II of two different genetic disorders: combined pituitary hormone deficiency and distal renal tubular acidosis could explain most of the complex clinical profile of this king, including his impotence/infertility which in the last instance led to the extinction of the dynasty.

The most famous genetic disease that circulated among European royalty was hemophilia. Because the progenitor, Queen Victoria, was in a first cousin marriage, it is often mistakenly believed that the cause was consanguinity. However, this disease is generally not aggravated by cousin marriages, although rare cases of hemophilia in girls (though not including Victoria) are thought to result from the union of hemophilic men and their cousins.

Intermarriage within European royal families has declined in relation to the past. Inter-nobility marriage was used as a method of forming political alliances among elite power-brokers. These ties were often sealed only upon the birth of progeny within the arranged marriage. Thus marriage was seen as a union of lines of nobility, not of a contract between individuals as it is seen today. Some Peruvian Sapa Incas married their sisters; in such cases a special combination between endogamy and polygamy is found. Normally the son of the old ruler and the ruler's oldest (half-)sister became the new ruler. The Inca had an unwritten rule that the new ruler must be a son of the Inca and his wife and sister. He then had to marry his sister (not half-sister), which ultimately led to the catastrophic Huáscar's reign, culminating in a civil war and then fall of the empire.

The Chakri Dynasty of Thailand has included marriages between cousins as well as more close relatives. The current king, Bhumibol Adulyadej is a first-cousin once removed of his wife, Sirikit, the two being respectively a grandson and a great-granddaughter of Chulalongkorn. The parents of the king's father, Mahidol Adulyadej, were half-siblings, both being children of Mongkut by different mothers.

Isolated Groups

Among genetic populations that are isolated, opportunities for exogamy are reduced. Isolation may be geographical, leading to inbreeding among people in remote mountain valleys. Or isolation may be social, induced by the lack of appropriate partners, such as Protestant princesses for Protestant royal heirs, in which case inbreeding is desired. Since the late Middle Ages, it is the urban middle class that has had the widest opportunity for outbreeding and the least desire to inbreed.

Some Inbreeding May Enhance Fertility Rate

A recent study in Iceland by the deCODE genetics company, published by the journal Science, found that third cousins produced more children and grandchildren, suggesting that "in spite of the fact that bringing together two alleles of a recessive trait may be bad, there is clearly some biological wisdom in the union of relatively closely related people.". For hundreds of years, inbreeding was historically unavoidable in Iceland due to its then tiny and isolated population.

South African Canine Breed Registry, SACBR

The SACBR, The Elite all Purebred Canine Registry. a Dog studbook that upholds all purebred Pedigree Canines ancestral and progeny records. Dog Registration for the responsible Canine breeder and Dog owner in Africa.Founded for the Dog Lover, that really cares about the welfare and breeding of purebred Canines, promoting healthy dogs. Registered dogs can bring you a pride of ownership and more valuable pets and companions. Owners of purebred Canines can benefit themselves by registering their bloodlines and progenies, knowing that their dogs belongs to a all dog Studbook that protects all purebreds for future generations.

Dual Registration for your Purebred Dog(s)

If your dog is already registered with another Canine registry that is not giving you the services or even customer satisfaction you need, Please apply for a Single dog Registration for every individual. Your dog keeps his or her current registry; they just gain all the benefits of being an SACBR Elite registered dog.

Single Dog Registration

A copy of your dog(s) pedigree or copies of their parent's pedigree from other Registries and Kennel clubs are all-acceptable for registration. Use a single Registration application form for each dog.

Litter Registration

Both Parents must be registered with the SACBR prior or simultaneously. The SACBR Code Of Ethics signed prior to Litter registration. "down-loadable" From this website only. Obligatory, Micro-chipping of puppies before the age of six month is not obligatory at the SACBR.

Registration Process

Complete the required SACBR Documents, obtainable from us in PDF format in our download page, Word documents also available via e-mail. We can Fax & Post your application form for you free of charge. All litter and Pedigrees and Kennel Certificates are printed (hard copy) and posted, email of documentations done only by request. (Not compulsory).

New Services

All Litter Birth registration Certificates and Pedigrees CAN BE e-mailed for your convenience to you, no more postal fees, only by request, hard copies posted applies if e mail not requested.

Breeding Methods

As a hobby breeder, breeding rats to involves a combination of breeding methods. Starting with high quality rats and using inbreeding, line breeding and occasional outcrosses will help a breeder to reach the goal of improvement in the line. Bettering the breed should encompass several factors: health, temperament, and conformation. The three methods used to selectively breed towards this goal are inbreeding, line breeding, and outcrossing. Tracking the offspring and future generations of any line (or outcross) provides essential data.

Definition of a Line

A line does not happen over night, it can take years and involves multiple inbred and line bred generations.It seems a common practice in some parts of the fancy to outcross repeatedly (generation after generation) without testing for health issues, and without stopping to evaluate the rats, without attempting to set desirable traits, etc. This practice does not meet the definition of having a line(s) and should not be referred to as such. Outcrossing can be used to strengthen or add a trait to a line, but it is the inbreeding of those offspring back to the foundation line that helps it to remain true to the original definition of a linse. Animals outcrossed whose offspring are bred to the outcrossed animals would become a related or sub line, but would denote a new line or "branch."

Line Breeding

Line breeding is a term that breeders use to denote a family of inbred rats that begins with a single or pair of foundation rats. Rat's within a line will have the same (or closely) lineage. Line breeding is accomplished by tightly inbreeding as well as by breeding rats that are less closely related (aunt/nephew, uncle/niece, cousin/cousin, grandparents/grandchildren). Linebreeding is used to set certain traits as well as to eliminate negative ones, therefore it is important not to breed two animals together that have the same fault.

One must also not breed related rats together just because of the pedigree, not all the rats from each litter are the same genetically. Considering health traits, temperament, and physical features will enable a breeder to choose the correct rats for each other, not just because they are related or because of sentimental attachment.

Inbreeding Tolerance

Different species have different degrees of inbreeding tolerance. And there are situations, such as when dealing with endangered species, where close inbreeding can be disastrous. In some mammal populations a high inbreeding coefficient can lead to inbreeding depression. And, it is not always about the accumulation of ?bad? traits. It is possible for the natural percentage of certain ?lethal? genes that each organism carries with no adverse effects to accumulate and cause problems such as immune deficiencies and fertility issues. With rats, the safe inbreeding quotient is quite high. In laboratories a line is not even considered inbreed until the 20th generation. Inbred lab strains are often achieved by breeding brother to sister in each generation. It is important, in research, to use healthy animals that are as genetically similar as possible so that test results show consistent data.

Outcrossing

Outcrossing is done to introduce new traits that are missing from a line. Dominate genes, such as rex, will exhibit in the first generation.

Recessive genes such as dumbo ear or a colour dilution will typically show in the second generation if the offspring from the outcross (who now carry but do not exhibit the trait) are breed back to the parent used as an outcross (who exhibits the trait and is therefore homozygous for the trait) or if the siblings are bred together.

Type & Outcrossing

Altering Colour

The coat colour of a rat can be altered by breeding it to a rat of a different colour and then back breeding to the outcrossed parent.

Breeding to a rat with an eye dilution can also change the colour in a line. On the other hand, strengthening a colour can be achieved by breeding to rats carrying less dilutions.

Modifying Physical Structure

Outcrossing can be a used as a tool to improve eyes or ears (shape, placement, or size). It can also be used in hopes of improving the size, shape, and/or length of the body, the head, or the tail.

Adding or Modifying Patterns or Markings

Breeding to marked or patterned rats can enable you to add or improve the trait in your line.

Temperament and Outcrossing

Little is known on how well temperament can be improved by outcrossing. And it can sometimes be difficult to determine if temperament issues are environmental or genetic. Generally the safest action is to avoid breeding any animal with a poor temperament, particularly if the rat shows aggression.

Health Issues & Outcrossing

Outcrossing can be used to improve vigour in a line that has been inbred or linebred for many generations if the litters appear to become consistently smaller or the pups no longer have the strong health (vigour) or size that is normally seen in that line.

It can also be done in an effort to "breed out" a particular medical trait, such as a tendency toward tumors, diabetes, or in a line exhibiting a genetic defect such as megacolon. Considering that a great many of these issues are rooted in genetics, breeding an outcross does not always eliminate a problem. It can mask it and perpetuate the issue.

Outcrossing may eliminate a health issue over time if the correct animals are chosen for breeding: the ones that somehow did not get the recessive(s) as well as outcrossed rats that also aren't carrying the problematic gene.

All too often, outcrossing to get rid of a problem just hides the recessives enabling the negative trait to show up later in the line (and in other lines as the rats are outcrossed more).

A good example of well meaning, but not necessarily well thought out, "avoidance" outcrossing involves the breeding of hairless lines that have lactation issues. Breeding to haired females who carry hairless will ensure that the offspring between them and the hairless sire will be fed. Unfortunately, there is a good chance that many of the offspring will still carry the genes that contribute to lack of lactation.

Some issues are connected to certain traits, as is the case in "high white rats" who are associated with aganglionic megacolon. Breeding away from the trait can only be accomplished by breeding away from the markings.

Hobby Breeder Considerations-Outcrossing

When doing an outcross, think very carefully about which animal to bring in, consider the reason why you are choosing to bring this particular rat in and if the rat will complement the rats you are working with. You want to improve those few traits with the outcross, but you do not want to ruin the work you have already done. With each positive trait the new rat also will be bringing in every problem and weakness from its own line and adding it to yours.

Hints

- Know the history of the outcross (health, temperament, and genetic)
- Have a goal in mind when choosing an outcross.
- Breed only to a few rats from your existing line(s).
- Test breed after an outcross for negative issues by inbreeding siblings or back breeding to parents
- Track the offspring placed with other breeders or owners
- Let the offspring mature as long as possible before breeding to give yourself more time to evaluate health and temperament
- Wait to breed the outcrossed rats back into your original line particularly if the outcross has a little known or unknown history.

Considering the potential that each outcross could bring in many problems over the next several generations that it may not be worth it to take the chance ruining your years of work. It is wise to maintain a portion of the line separate from outcrosses to ensure the perpetuation of the healthy line in the event an outcross has negative results. You do not want to bring in something that is going to give you a whole new problem to "breed out."

Strain

A strain is a variation of a particular species that possesses minor differences in its characteristics (physiological or chemical) though still remain distinguishable. A strain requires homozygosity through close inbreeding methods such as brother/sister mating or by back crossing offspring with parents. Homozygosity is the presence of the same alleles at one or more loci, it is "genetic sameness."

The word strain comes from the Middle English word *streen*: meaning progeny, lineage, as well as from the Old English *streon*:

meaning gain, acquisition. Strains in the laboratory can include: inbred, outcrossed, and sub strains. In the lab it takes 20 generations of inbreeding to produce a strain that will be 95% genetically similar. With 40 inbred generations the percentage reaches 99.5%

Origin and Use

Rats were first used in Europe for nutritional studies as early as 1850. As time passed rats were kept in labs and used for other testing. The specific inbreeding of rats was primarily achieved in America. The oldest known strains of inbred rats were started by Helen Dean King in 1909 at the Wistar institute in Pennsylvannia. She had two lines of albino rats, one of which became known as the King Albino (Later named the PA strain). By 1920 she had reached the 135th generation of inbreeding the PA line. King also started an inbred strain from wild Norway rats that she caught locally. When it reached generation 35 it was designated the BN (Brown Norway) strain. Another scientist at Wistar, Margaret Dean Lewis, was the initiator of the LEW (Lewis) strain which reached its 8th generation in 1956. These strains were the origination either in part/or in whole of many of the modern day rat strains. The direct descendants of the Wistar strains are still in use today.

Use

The fact that the rat's physiology as well as its short life span have made it the lab animal most responsible for advances in medicine. Having such closely related strains in the laboratory is important to providing consistent data to researchers. It provide lines of rats with no apparent health or disease issues as well as rats with specific genetic issues. The rats with no apparent genetic issues help the researchers more about such things as pathogenic disease, neurological injuries, toxicology, behaviour, and pharmacology. Rats with specific genetic issues are used by comparative medicine researchers to learn more about disease process and genetic disorders, behaviour, treatment of diseases, causes of disorders, and pharmacology. Comparative medicine research involving rats, although hard for rat lovers to sometimes deal with, has enabled medical doctors to learn more about the treatment and prevention of many things that people suffer from such as cancer, diabetes, birth defects, and many others.

2

Animal Breeding Methods

The following animal breeding methods are followed for improvement of animals in dairy and poultry namely

Inbreeding

The crossing of closely related animals is called Inbreeding. If this inbreeding is repeated continuously, it is called Upgrading. Inbreeding is used to retain desirable genetic traits in animals. Inbreeding is required in order to retain as many traits as possible by keeping the combination of genes intact.

However, the inbreeding may result in homozygous recessive genes coming together to express some harmful phenotypic traits. Many breeding scientists have observed that hybrid vigour and fertility were lost due to repeated inbreeding. Such recessive and harmful genes are removed by some special techniques without sacrificing the major quality of the animal. If the race is relatively free of such harmful recessive genes, the process of inbreeding is a safe method for improvement of animals.

Outbreeding

The crossing of distantly related animals is called Outbreeding. One of the problems the animal breeder faces in outbreeding is introduction of new genes into population. In this method it is possible to breed a desirable type of animal with a less desirable type and then to increase the degree of desirable traits. New and high yielding genes can be introduced into the population through outbreeding. In many cases these genes may come from a variety of stock. Out breeding in animals is useful for different purposes viz. To produce some valuable traits ii. To create new breeds iii. To produce a hybrid of superior vigour and value. To produce some valuable traits: Beef cattle may be crossed with dairy cattle to produce calves for superior veal (flesh) production.

To create new breeds: A new breed is produced with desired characters from the two original breeds. This process of producing new breeds takes time. The present day breeds of animals have been developed through hybridisation.

To produce a Hybrid of Superior Vigour and Value

A Mule is produced by crossing Equus equus (mare or female horse) and Equus homonius (jack or male ass). Mules are superior to horses in strength, endurance, resistance to disease and ability to work under unfavorable conditions. When a female mule is crossed with a jack, a colt is produced.

Mutations

The new traits into populations can also be induced through mutations. Since most of the mutations are harmful and the process of induction of mutations is quite expensive, this method of improvement of animals is impractical. It has been reported that a sheep in New England mutated in the direction of having shorter legs (a desirable quality) and formed the basis for racial improvement of sheep.

Representative examples of improvements in animals by mutations include poultry that is resistant to white diarrhoea, increased egg production in fowls, increased fat content in milk and better meat yielding in Turkeys etc.

Animal Breeding

Animal breeding is the selective mating of animals to increase the possibility of obtaining desired traits in the offspring. It has been performed with most domesticated animals, especially cats and dogs, but its main use has been to breed better agricultural stock. The more modern techniques involve a wide variety of laboratory methods, including the modification of embryos, sex selection, and genetic engineering. These procedures are beginning to supplant traditional breeding methods, which focus on selectively combining and isolating livestock strains. In general, the most effective strategy for isolating traits is by selective inbreeding; but different strains are sometimes crossed to take advantage of hybrid vigour and to forestall the negative results of inbreeding, which include reduced fertility, low immunity, and the development of genetic abnormalities.

The Genetic Basis of Animal Breeding

Breeders engage in genetic "experiments" each time they plan a mating. The type of mating selected depends on the goals. To some breeders, determining which traits will appear in the offspring of a mating is like rolling the dice—a combination of luck and chance. For others, producing certain traits involves more skill than luck—the result of careful study and planning....

Economic Considerations

There are many reasons why animal breeding is of paramount importance to those who use animals for their livelihood. Cats have been bred largely for aesthetic beauty; many people are willing to pay a great deal of money for a Siamese or Persian cat, even though the affection felt for a pet has little to do with physical appearance. But the most extensive animal breeding has occurred in those areas.

Modern Methods in Biotechnology

Artificial insemination is the artificial introduction of semen from a male with desirable traits into females of the species to produce pregnancy, and is useful because a far larger number of offspring can be produced than would be possible if the animals were traditionally bred. Because of this, the value of the male as breeding stock can be determined much more rapidly, and the use of many diff...

Embryo Manipulation

In order to understand the techniques of embryo manipulation, it is important to understand the early stages of reproduction. When the egg and sperm unite to form a zygote, each of the parents supply the zygote with half of the chromosomes necessary for a full set. The zygote, which is a single cell, then begins to reproduce itself by the cellular division process called mitosis.

Genetic Engineering

Genetic engineering is being implemented to create animals that have had a new gene inserted directly into their DNA. These animals are called transgenic. The procedure involves microinjection of the desired gene into the nucleus of fertilized eggs. It has been found that in many cases, but with varying rates of success, the new gene is reproduced in all developing cells, and the gene can be trans...

Sex Selection

It would be extremely useful if a breeder were able to predetermine the sex of each embryo produced, because in many cases one sex is preferred. For instance, in a herd of dairy cows or a flock of laying hens, females are the only commercially useful sex. When the owner of a dairy herd has inseminated a cow at some expense, this issue becomes more crucial. In some cases, an animal is being bred.

Animal breeding, controlled propagation of domestic animals in order to improve desirable qualities. Humanity has been modifying domesticated animals to better suit human needs for centuries. Selective breeding involves using knowledge from several branches of science. These include genetics, statistics, reproductive physiology, computer science, and molecular genetics. This article discusses the basic principles of how populations of animals can be changed by application of these principles, and a brief discussion of molecular genetics, immunogenetics, and newer reproductive technologies is included. The fundamental biological principles underlying animal breeding are discussed in the articles heredity and animal reproductive system.

Breeding and Variation

English agriculturist Robert Bakewell was a very successful breeder of commercial livestock in the 18th century. His work was based on the traditional method of visual appraisal of the animals that he selected. Although he did not write about his methods, it is recorded that he travelled extensively by horseback and collected sheep and cattle that he considered useful. It is thought that he made wide outcrosses of diverse breeds, and then practiced inbreeding with the intent of fixing desirable characteristics in the crossbred animals. He was also the first to systematically let his animals for stud. For these reasons he is generally recognized as the first scientific breeder.

In animal breeding, a population is a group of interbreeding individuals—i.e., a breed or strain within a breed that is different in some aspects from other breeds or strains. Typically, certain animals within a breed are designated as purebred. The essential difference between purebred and nonpurebred animals is that the genealogy of purebred animals has been carefully recorded, usually in a herd book, or studbook, kept by some sanctioning association. Purebred associations provide other services that are useful to their members to enhance their businesses.

Selective breeding utilizes the natural variations in traits that exist among members of any population. Breeding progress requires understanding the two sources of variation: genetics and environment. For some traits there is an interaction of genetics and the environment. Differences in the animals' environment, such as amount of feed, care, and even the weather, may have an impact on their growth, reproduction, and productivity. Such variations in performance because of the environment are not transmitted to the next generation.

Genetic variation is necessary in order to make progress in breeding successive generations. Each gene, which is the basic unit of heredity, occupies a specific location, or locus, on a chromosome. Two or more genes may be associated with a specific locus and therefore with a specific trait. (Traits that can be observed directly, such as size, colour, shape, and so forth, make up an organism's phenotype.) These genes are known as alleles. If paired alleles are the same, the organism is called homozygous for that trait; if they are different, the organism is heterozygous. Typically, one of the alleles will be expressed to the exclusion of the other allele, in which case the two alleles are referred to as dominant and recessive, respectively. However, sometimes neither dominates, in which case the two alleles are called codominant.

Although no complete knowledge of the genetic makeup of any breed of livestock exists yet, genetic variations can be used for improving stock. Researchers partition total genetic variation into additive, dominance, and epistatic types of gene action, which are defined in the following paragraphs. Additive variation is easiest to use in breeding because it is common and the effect of each allele at a locus just adds to the effect of other alleles at that same locus. Genetic gains made using additive genetic effects are permanent and cumulate from one generation to the next.

Although dominance variation is not more complex in theory, it is more difficult to control in practice because of how one allele masks the effect of another. For example, let a indicate a locus, with a_1 and a_2 representing two possible alleles at that location. Then a_1a_1, a_1a_2 (which is identical to a_2a_1), and a_2a_2 are the three possible genotypes. If a_1 dominates a_2, the genotypes a_1a_2 and a_1a_1 cannot be outwardly distinguished. Thus, the inability to observe differences between a_1a_2

and a_1a_1 presents a major difficulty in using dominance variance in selective breeding.

Breeding Objectives

Breeding objectives can be discussed in terms of changing the genetic makeup of a population of animals, where population is defined as a recognized breed. Choice of breeding goals and design of an effective breeding program is usually not an easy task. Complicating the implementation of a breeding program is the number of generations needed to reach the initial goals. Ultimately, breeding goals are dictated by market demand; however, it is not easy to predict what consumers will want several years in advance. Sometimes the marketplace demands a different product than was defined as desirable in the original breeding objective. When this happens, breeders have to adjust their program, which results in less-efficient selection than if the new breeding goal had been used from the beginning. For example, consumers want leaner beef that is tender. Thus, ranchers have changed their cattle-breeding programs to meet this new demand. These trends have gradually changed over the last few decades; for example, Angus cattle are particularly noted for the quality of beef produced. The use of ultrasound is now widespread in determining the fat and lean content of live animals, which will hasten the changing of carcass quality to meet consumer demands.

Additional complications arise from simultaneously trying to improve multiple traits and the difficulty of determining what part of the variation for each trait is under genetic control. In addition, some traits are genetically correlated, and this correlation may be positive or negative; that is, the traits may be complementary or antagonistic. Breeding methods depend on heritability and genetic correlations for desirable traits.

Heritability and Genetic Correlations in Breeding

Heritability is the proportion of the additive genetic variation to the total variation. Heritability is important because without genetic variation there can be no genetic change in the population. Alternatively, if heritability is high, genetic change can be quite rapid, and simple means of selection are all that is needed. Using an increasing scale from 0 to 1, a heritability of 0.75 means that 75 percent of the total variance in a trait is controlled by additive gene action. With heritabilities this high, just the record of a single individual's traits can easily be used to create an effective breeding program. Some general statements can be made about heritability, keeping in mind that exceptions exist. Traits related to fertility have low heritabilities. Traits related to production have intermediate heritabilities. Examples include the amount of milk a cow produces, the rates of weight gain in steers and pigs, and the number of eggs laid by chickens. So-called quality traits tend to have higher heritabilities. Examples include the amount of fat a pig has over its back and the amount of protein in a cow's milk. The magnitude of heritability is one of the primary considerations in designing breeding programs.

Genetic correlation occurs when a single gene affects two traits. There may be many such genes that affect two or more traits. Genetic correlations can be positive or negative, which is indicated by assigning a number in the range from +1 to " 1, with 0 indicating no genetic correlation. A correlation of +1 means that the traits always occur together, while a correlation of " 1 means that having either trait always excludes having the other trait. Thus, the greater the displacement of the value from 0, the greater the correlation (positive or negative) between traits. The practical breeding consequence is that selection for one trait will pull along any positively correlated traits, even though there is no deliberate selection for them. For example, selecting for increased milk production also increases protein production. Another example is the selection for increased weight gain in broiler chickens, which also increases the fat content of the birds.

When traits have a negative genetic correlation, it is difficult to select simultaneously for both traits. For example, as milk production is increased in dairy cows through genetic selection, it is slightly more difficult for the high-producing cows to conceive. This negative correlation is partly due to the partitioning of the cows' nutrients between production and reproduction, with production being prioritized in early lactation. In the case of dairy cattle, milk production is on the order of 20,000 pounds per year and is increasing. This is a large metabolic demand, so nutrient demand is large to meet this need. Thus, selecting for improved fertility may result in a reduction in milk production or its rate of gain.

Methods of Selection

Types of selection are individual or mass selection, within and between family selection, sibling selection, and progeny testing, with many variations. Within family selection uses the best individual from each family for breeding. Between family selection uses the whole family for selection. Mass selection is most effective when heritability is high and the trait is expressed early in life, in which case all that is required is observation and selection based on phenotypes. When mass selection is not appropriate, other methods of selection, which make use of relatives or progeny, can be used singularly or in combination. Modern technologies allow use of all these types of selection at the same time, which results in greater accuracy.

Elements Needed to Make Genetic Progress

Genetic gain per year (ΔG) depends on balancing several factors, as expressed in the equation $\Delta G=(A\sigma_g i)/I$, where A is the accuracy of selection, σ_g is the standard deviation of the additive genetic variation in the population, i is the selection intensity (proportion selected for further breeding), and I is the generation interval (age of breeding). The σ_g factor cannot be easily changed within a breed, though it can be changed by crossbreeding. The other quantities in the equation can be changed. More complete pedigree records on candidates for breeding can increase the accuracy of selection, but waiting for candidates to reach full maturity in order to have better genetic data will increase the generation interval. Whether an increase in generation interval is justified by a more accurate selection process depends on individual circumstances. Selection intensity can also be increased, by narrowing the proportion of the population used in breeding, but it should be done without increasing the generation interval. Because generation interval is the divisor in the genetic gain equation, anything that increases the generation interval has an unfavourable impact on genetic progress, all else being equal.

Evaluation of Animals

Methods of ranking animals for breeding purposes have changed as statistical and genetic knowledge has increased. Along with increases in breeding knowledge, advancements in computing have enabled breeders to quickly and easily process routine breeding evaluations, as well as to develop research needed to rank large populations of animals. Evaluating and ranking candidates for selection depends on equating their performance record to a statistical model. A performance record (y) can be expressed as $y = g + e + \varepsilon$, where g stands for genetic effects, e indicates known (categorized) environmental effects, and ε indicates random environmental effects.

The first task in estimating *g* is to statistically eliminate environmental effects, a process that involves setting up a system of equations to simultaneously solve for all of the genetic effects for the sires and cows. Information from relatives is included in *g* and increases the accuracy of evaluation. All relatives that are available can be incorporated in this type of evaluation. This model is called the animal model.

The animal model is used extensively in evaluating beef and dairy cattle, chickens, and pigs. To apply this model for evaluating large populations requires use of high-speed computers and extensive use of advanced mathematical techniques from numerical analysis. In evaluating the dairy cattle in the United States, a system of equations with more than 25 million variables is needed.

Accuracy of Selection

In some cases the accuracy of selection for a trait can be measured using a calibrated tool or a scale. Thus, measurements of such traits can be replicated with high reliability. Alternatively, some traits are difficult to measure on an objective scale, in which case a well-designed subjective scoring method can be effective. An excellent example is hip dysplasia, a degenerative disease of the hip joints that is common in many large dog breeds. Apparently, hip dysplasia is not associated with a single allele, making its incidence very difficult to control. However, an index has been developed by radiologists that allows young dogs to be assigned a score indicating their likelihood of developing the disease as they age. In 1997 American animal geneticist E.A. Leighton reported that, in fewer than five generations of selection in a breeding experiment using these scores, the incidence of canine hip dysplasia in German shepherd dogs measured at 12 to 16 months of age had decreased from the breed average of 55 percent to 24 percent among the experimental population; in Labrador retrievers the incidence dropped from 30 to 10 percent.

Because close relatives share many genes, an examination of the relatives of a candidate for breeding can improve accuracy of selection. The more complete the genealogical record, or pedigree, the more effective the selection process. A pedigree is most useful when the heritability of a trait is relatively low, especially for traits that are expressed later in life or in only one sex.

Reproductive techniques can be used to increase the rate of genetic progress. In particular, for species that are mostly bred by artificial insemination, the best dams can be chosen and induced to superovulate, or release multiple eggs from their ovaries. These eggs are fertilized in the uterus and then flushed out in a nonsurgical procedure that does not impair future conception of the donor female. Each embryo is implanted in a less-valuable host female to be carried through gestation. The sex of the embryos can be determined in utero at about 50 days of gestation. The normal gestation for Holstein-Friesian cattle is about 280 days, so this early determination of sex saves many days and allows the breeding program to be adjusted. In particular, the donor cow could be collected again, or another superior cow could be bred to produce males. Thus, these reproductive technologies reduce the generation interval and increase selection intensity by getting more than one male calf from superior females. Both superovulation and sex determination are now commonly used procedures. Superovulation is also used when breeders want to increase the number of female calves from a valuable cow.

Progeny Testing

Progeny testing is used extensively in the beef and dairy cattle industry to aid in evaluating and selecting stock to be bred. Progeny testing is most useful when a high level of accuracy is needed for selecting a sire to be used extensively in artificial insemination. Progeny testing programs consist of choosing the best sires and dams in the population based on an animal model evaluation, as described in the preceding section. A description of progeny testing in dairy breeding provides a good example. The best 1 to 2 percent of the cows from the population are chosen as bull mothers, and the best progeny-tested bulls are chosen to produce another generation of sires. The parents are mated to complement any individual deficiencies. The accuracy of evaluation of bull mothers is typically about 40 percent, and of sires that produce young bulls the accuracy is more than 80 percent. This is not as high as the industry wants for bulls to be used in artificial insemination. To reach greater accuracy, the next generation of sires is mated to enough cows in the population for each sire to produce 60 to 80 progeny. After the daughters of the young sires have a production record, the young sires are evaluated, and about the best 10 percent are used extensively to produce commercial cows. Some of the progeny-tested sires will have thousands of daughters before a superior sire is found to replace them. About 70 percent of dairy cattle are bred by artificial insemination, so these sires control the genetic destiny of dairy cattle. The genetic gain has been consistent over the years. The actual first-lactation milk production varies more than the sire breeding value because differences in environmental conditions affect first-lactation production, but these environmental effects have been adjusted out of the breeding value calculations. There is no indication that the rate of gain in the sire breeding values is about to reduce. This level of achievement can only be realized if artificial insemination organizations and producers work together.

Crossbreeding

Crossbreeding involves the mating of animals from two breeds. Normally, breeds are chosen that have complementary traits that will enhance the offsprings' economic value. An example is the crossbreeding of Yorkshire and Duroc breeds of pigs. Yorkshires have acceptable rates of gain in muscle mass and produce large litters, and Durocs are very muscular and have other acceptable traits, so these breeds are complementary. Another example is Angus and Charolais beef cattle. Angus produce high-quality beef and Charolais are especially large, so crossbreeding produces an animal with acceptable quality and size.

The other consideration in crossbreeding is heterosis, or hybrid vigour, which is displayed when the offspring performance exceeds the average performance of the parent breeds. This is a common phenomenon in which increased size, growth rate, and fertility are displayed by crossbred offspring, especially when the breeds are more genetically dissimilar. Such increases generally do not increase in successive generations of crossbred stock, so purebred lines must be retained for crossbreeding and for continual improvement in the parent breeds. In general, there is more heterosis for traits with low heritability. In particular, heterosis is thought to be associated with the collective action of many genes having small effects individually but large effects cumulatively. Because of hybrid vigour, a high proportion of commercial pork and beef come from crossbred animals.

Inbreeding

Mating animals that are related causes inbreeding. Inbreeding is often described as "narrowing the genetic base" because the mating of related animals results in offspring that have more genes in common. Inbreeding is used to concentrate desirable traits. Mild inbreeding has been used in some breeds of dogs and has been extensively used in laboratory mice and rats. For example, mice have been bred to be highly sensitive to compounds that might be detrimental or useful to humans. These mice are highly inbred so that researchers can obtain the same response with replicated treatments.

As a result, producers try to avoid mating related animals. This is not always possible, though, when long-continued selection for the same traits is practiced within a small population, because parents of future generations are the best candidates from the last generation, and some inbreeding tends to accumulate. The rate of inbreeding can be reduced, but, if inbreeding depression becomes evident, some method of introducing more diverse genes will be needed. The most common method is some form of crossbreeding.

The Genetic Code and Immunogenetics DNA

Deoxyribonucleic acid (DNA) is the genetic material that contains the instructions in each cell of organisms. DNA determines the genome, and thus the genetic code, which is a blueprint for development of all body organs and structures. The structure of DNA can be visualized as a spiral staircase. The handrails are made up of sugar and phosphate molecules, and the steps are composed of four nitrogenous bases: adenine, thymine, cytosine, and guanine. These bases are paired: adenine is paired with thymine, and cytosine is paired with guanine. The order of these four base pairs is the genetic code that determines the genotype of an individual. The DNA is arranged on chromosomes inside cells, with cells having two methods of dividing and replicating. In mitosis, a cell divides into two daughter cells such that each contains an exact copy of the original cell's chromosomes. In meiosis, a germ cell's chromosomes are duplicated before the cell undergoes two divisions to produce four gametes, or sex cells, each with half (male or female) of the original cell's chromosomes. During the process of fertilization, male and female gametes from different organisms pair their chromosomes to form a zygote, which eventually becomes an adult.

Genetic progress in domestic animals has been made using quantitative methods to date. It would be very desirable to know the genes that control the many traits that have economic significance in domestic animals. This should make selection more accurate. Information from sequencing human genes, as well as those of other species, is being used to find chromosomal segments with high probabilities of coding genes in livestock. Another approach is to scan a chromosome segment and look for associations with economic traits. Several quantitative trait loci have been discovered that are or promise to be useful in livestock breeding. For example, an estrogen receptor in pigs is associated. Genes are also known for growth hormone, and many others could be enumerated. With improvements in sequencing DNA, more genes will be discovered that affect economic traits— genes that will need to be tested in different genetic backgrounds and environments before they can be commercialized.

It is now much less expensive to sequence DNA, which has led to new methods of evaluating animals using large segments of 30,000– 50,000 bases. With the use of these large segments of DNA, animals are evaluated without looking for markers for individual traits. This is intuitively an appealing approach because much more of the DNA can be evaluated; perhaps in the future the entire genome can be used to evaluate animals. This method of selection, called genomic selection, is now being applied to dairy cattle, but results are not yet available.

Immunogenetics

The connection between an organism's genetic makeup and its immune system, as well as applications of that knowledge, form the young science of immunogenetics. In particular, producers must control diseases in their livestock if they are going to be profitable. While vaccines, hygiene, and other therapeutic methods control most diseases, vaccines are expensive and none of these methods is completely effective. However, there is evidence from experiments and field data of some degree of genetic control over the immune system in humans and animals. For example, bovine leukocyte adhesion deficiency (BLAD) is a hereditary disease that was discovered in Holstein calves in the 1980s. The presence of the BLAD gene leads to high rates of bacterial infections, pneumonia, diarrhea, and typically death by age four months in cattle, and those that survive their youth have stunted growth and continued susceptibility to infections. It was soon found that these calves carried two copies of a recessive gene that was present in nearly 25 percent of Holstein bulls. Cattle with only one copy of the gene, or carriers, had normal growth patterns and immune systems. Holstein bulls are now routinely tested for the BLAD gene before being used for artificial insemination. With a high percentage of Holsteins being bred artificially, a potentially major problem has been avoided.

Histocompatibility genes that serve several functions are on one area of a chromosome, called the major histocompatibility complex

(MHC), which exists in all higher vertebrates. There are large numbers of genes involved in the MHCs of different species. There are more than 60 different alleles at one locus and other loci are multi-allelic. There are also differences among species in the number of genes known. In addition, selection experiments have demonstrated genetic variation between lines selected for high and low response to different antigens. Some vaccinations are more efficacious when the animals have been selected for resistance to the antigen for which they are vaccinated.

Substantial progress has been made in the field of immunogenetics, but limited use has been made of this knowledge. One reason for this is that immune systems have evolved to be generally robust. Changing the frequency of some genes that control immune function may inadvertently change the function of other genes and result in adverse effects. Experiments are now under way to determine whether sires' immune responses can be used to predict the health of their daughters under field conditions. The results indicate that there are differences among sires' daughter groups, but the differences are not large enough to control a high proportion of the variability. The tests used were based primarily on leukocytes, which are the first line of defense when an antigen invades an animal. Application of knowledge in the area of immunogenetics must be used with caution.

It might seem that integrating molecular markers and quantitative methods would be a trivial task. However, the effect of some genes depends on the presence of others, and these interactions need to be considered along with the particular breeding scheme. Furthermore, there are nongenetic influences that may turn genes on and off. Thus, some genes act individually, some genes interact, and the environment has a further impact. Finding how these all affect the phenotypic expression of an organism is complicated. However, this challenge presents an opportunity for future research and for producers.

Many advances in reproductive technologies have been made, though many are too expensive for everyday use. Most of the advanced techniques use artificial insemination, which was developed decades ago, though refinements continue.

Cloning

Cloning, an asexual method of reproduction, produces an individual with the same genetic material (DNA) as another individual. Probably the best-known examples of Though the DNA in cloned individuals is the same, environmental influences may make them differ in phenotype. Thus far, the commercial use of clones has been limited. Cloning can be used to produce clones from a highly productive individual, but the cost would have to be low enough to recover the expense quickly. Animals have been cloned by three processes: embryo splitting, blastomere dispersal, and nuclear transfer. Nuclear transfer is most common and involves enucleating an ovum, or egg, with all the genetic material removed. This material is replaced with a full set of chromosomes from a suitable donor cell, which is microinjected into the enucleated cell. Then the enucleated cell, with the transplanted chromosome, is placed into a recipient female to be carried through gestation.

Determining Sex From Sperm

There is a commercial demand for the ability to predetermine the sex of livestock. For example, a producer may want female calves from the best cows for replacements and male calves for beef production. Dairy producers may want more females for replacing cows or for expansion of their herds. The sex of mammals is determined by the sex chromosomes, or X and Y chromosomes. Animals with two X chromosomes develop into females; animals with an X and a Y chromosome develop into males. Thus, the detection of X and Y chromosomes on sperm has been the focus of research to predetermine the sex of domestic animals.

In one process, sperm is pretreated with a dye that fluoresces when exposed to short wavelength light. The fluorescence is brighter from a sperm bearing X chromosomes, which contain about 4 percent more DNA than the Y chromosome. A stream of dyed sperm is passed through a flow cytometer, a computer determines the degree of fluorescence, and the sperm is separated into different containers.

The success rate can be as high as 40 percent. When "sexed sperm" has been used on a commercial basis, though, it has had limited success. The conception rate using sexed sperm is lower in cows, though it is higher in primiparous cows. In addition, sperm are killed in the typing process, and the rate of sexing the sperm is slower than desired. While economical processing of sperm is just getting started, it is expected to become another useful tool in animal agriculture.

Aspects of the topic animal breeding are discussed in the following places at Britannica.

Assorted References

Offspring that have purebreeding genotypes (i.e., *AA, bb, cc,* or *DD*). This type of experimental breeding is the origin of new plant and animal lines, which are an important part of making laboratory stocks for basic research. When applied to commerce, transgenic commercial lines produced experimentally are called genetically modified...

Application

The goal of animal breeders in the 20th century was to develop types of animals that will meet market demands, be productive under adverse climatic conditions, and be efficient in converting feed to animal products. At the same time, producers have increased meat production by improved range management, better feeding practices, and the eradication of diseases and harmful insects. The world.

By the 18th century, bullfighting's popularity had grown sufficiently to make bull breeding financially profitable, and herds were bred for specific characteristics. In fact, many of the royal houses of Europe competed to present the fiercest specimens in the ring. The lack of a spirited native stock of bulls is one reason why corridas never fully took root in Italy and France.

To satisfy this demand, sheep, goats, cattle, water buffalo, swine, chickens, ducks, geese, and turkeys are produced on farms all over the world. To understand how agricultural animals convert feedstuffs into the food and other commodities consumers demand, animal scientists have undertaken broad investigations using highly sophisticated techniques. The animal sciences...

The guiding principle for breeding winning racehorses has always been best expressed as "breed the best to the best and hope for the best." The performance of a breeding horse's progeny is the real test, but, for horses untried at stud, the qualifications are pedigree, racing ability, and physical conformation.

Arab and Barb horses was introduced into England as early as the 3rd century. Natural conditions favoured development of the original stock, and selective breeding was encouraged by those interested in racing. Under the reigns of James I and Charles I, 43 mares—the so-called Royal Mares—were imported into England, and a record, the General. Finally, pets themselves have become a self-perpetuating industry, bred for a variety of purposes, including their value as breeding animals. Pets that are bred for aesthetic purposes may have full-fledged show careers. Other pets may be bred for racing or other competitive sports, around which sizable industries have been built.

The British Isles led the world in the development of the principal beef breeds; Herefords, Angus, beef Shorthorns, and Galloways all originated in either England or Scotland. Other breeds of greatest prominence today originated in India (Brahman), France (Charolais; Limousin; Normandy), Switzerland (Simmental), and Africa (Africander). The Hereford breed, considered to be the first to be.

Even though all cats are similar in appearance, it is difficult to trace the ancestry of individual breeds. Since tabbylike markings appear in the drawings and mummies of ancient Egyptian cats, present-day tabbies may be descendants of the sacred cats of Egypt. The Abyssinian also resembles pictures and statues of Egyptian cats.

The good health are important criteria for choosing a cat. Disposition varies only slightly between male and female cats. There are, however, distinct differences in disposition among the various pedigreed varieties; the Siamese, for example, is vocal and demanding, while the Persian is quiet and fastidious. The mixed breed, or "alley" cat, is a heterogeneous breed of unknown.

Once it became evident that dogs were faster and stronger and could see and hear better than humans, those specimens exhibiting these qualities were interbred to enhance such attributes. Fleet-footed sight hounds were revered by noblemen in the Middle East, while in Europe powerful dogs such as the mastiff were developed to protect home and traveller from harm.

The northern species is nonmigratory; the southern elephant seal, like the northern form, breeds and molts on land, but it winters at sea, possibly near the pack ice. During the breeding season, elephant seals become aggressive toward each other. The bulls fight to establish territories along beaches and to acquire harems of up to 40 cows. The cows produce single brownish.

Fish farming as originally practiced involved capturing immature specimens and then raising them under optimal conditions in which they were well fed and protected from predators and competitors for light and space. It was not until 1733, however, that a German farmer successfully raised fish from eggs that he had artificially obtained and fertilized. Male and female trout were collected. Their selective breeding of ornamental goldfish was later introduced to Japan, where the breeding of ornamental carp was perfected. The ancient Romans, who kept fish for food and entertainment, were the first known marine aquarists; they constructed ponds that were supplied with fresh.

The first intensively domesticated horses were developed in Central Asia. They were small, lightweight, and stocky. In time, two general groups of horses emerged: the southerly Arab-Barb types (from the Barbary coast) and the northerly, so-called cold-blooded types. When, where, and how these horses appeared is disputed. Nevertheless, all modern breeds—the light, fast, spirited breeds.

Vandals. When these barbarian peoples invaded the empire, the vast number of horses that they possessed helped them to overthrow the Romans. The era that followed witnessed the collapse of the Roman breeds and the gradual development—especially during the era of Charlemagne in the late 8th and early 9th centuries—of improved types, owing largely to the importation of Arabian stock..

The breeds of dairy cattle have been established by years of careful selection and mating of animals to attain desired types. Increased milk and butterfat production has been the chief objective, although the objective often has shifted to increased milk and protein production. Production per cow varies with many environmental factors, but the genetic background of the cow is extremely.

The reserve system has been expanded from 14 sites to more than 40, and cooperative international arrangements were implemented to provide training in reserve management and captive breeding. Prior eras of giving pandas as gifts and of short-term commercial loans to zoos have given way to lending agreements that generate funds for preservation of the wild population.

Chicken breeding is an outstanding example of the application of basic genetic principles of inbreeding, linebreeding, and crossbreeding, as well as of intensive mass selection to effect faster and cheaper gains in broilers and maximum egg production for the egg-laying strains. Maximum use of heterosis, or hybrid vigour, through incrosses and crossbreeding has been made. Crossbreeding for egg...

Shepherdy and nomadic animal breeding, which determined the social and economic organization and the way of life of some peoples to a great extent, appeared at later stages of human development, after the accumulation of a large number of domestic animals. Rudiments of nomadic animal breeding in Eurasia appeared no earlier than 1000 bce, considerably after the domestication of animals took... One of the most important developments was the management of animal herds for purposes other than the provision of meat. In the case of cattle, there is some evidence for milk production earlier, but dairying appears to have taken on a much more significant role from this time. Oxen were raised to provide traction. Sheep were managed not for meat but primarily as a source of manure and wool.

Since World War II a number of zoos have been developed as breeding centres for animal species in danger of becoming extinct in the wild. Many threatened species have been saved by breeding in captivity. For example, in 1947 it was estimated that there were only 50 nenes, or Hawaiian geese, left on Hawaii and none anywhere else in the world. In 1950 two nenes were housed at the Wildfowl Trust...

Animal Breeding

The science of ANIMAL breeding is defined as the application of the principles of GENETICS and biometry to improve the efficiency of production in farm animals. These principles were applied to change animal populations thousands of years before the sciences of genetics and biometry were formally established. The practice of animal breeding dates back to the Neolithic period (approximately 7000 BC), when people attempted to domesticate wild species such as caribou, goats, hogs and DOGS.

Domestication was performed through controlled mating and reproduction of captive animals which were selected and mated based on their behaviour and temperament. Judging from cave paintings that have survived, selection was also applied to some qualitative traits such as coat colour and the absence or presence of horns. Without written records, there is no certain knowledge of the evolution of animal breeding practices, but written documents dating back more than 4000 years indicate that humans appreciated the significance of family resemblance in mating systems, recognized the dangers of intense inbreeding, and used castration to prevent the reproduction of undesirable males. Progress in the performance of domesticated animals through these selection practices was very slow; improvements were mainly due to animals adapting better to their environments.

Robert Bakewell, an English animal breeder of the 18th century, is considered the founder of systematized animal breeding. He was the first to emphasize the importance of accurate breeding records, introduced the concept of progeny testing to evaluate the genetic potentials of young sires, and applied inbreeding to stabilize desired qualitative traits. He also promoted concepts such as "like begets like,""prepotency is associated with inbreeding" and "breed the best to the best." Bakewell and his contemporaries in Europe pioneered the development of diverse breeds of BEEF cattle, DAIRY cattle, SHEEP, hogs and HORSES.

Most livestock breeds with pedigree herd books and breed associations were established between the late 18th century and the second half of the 19th century. Colour, conformation, geographical origin and some production characteristics were the main factors that differentiated these breeds. Wide geographical redistribution of animal populations was also an important factor in the formation of new breeds, as invading armies, migrating people and traders transported livestock to new lands.

Animal breeding as a modern SCIENCE belongs to the 20th century. Although numerous geneticists and biometricians have made significant contributions to the development of this science, J.L. Lush of Iowa State University is considered the father of the modern science of animal breeding. Lush and his students developed major scientific procedures applicable to the genetic improvement of farm animals.

Animal Breeding in Canada

Studies on crossbreeding were first performed at the University of Saskatchewan in 1930, under the direction of J.W. Grant MACEWAN and L.M.Winters. Studies on quantitative genetics in Canada were initiated by Jack Stothart (1934). Since 1940, Agriculture and Agri-Food Canada RESEARCH STATIONS at Lacombe and Lethbridge, Alta, Brandon, Man, Lennoxville, Qué, and Ottawa have been active in animal breeding research. Among educational centres, the universities of Guelph, Alberta, McGill and Manitoba have been active in animal breeding research and training. The relative scarcity of scientists and active research centres in this field reflects the high cost of research on genetic improvement of farm animals.

Major crossbreeding studies and breed synthesis projects have included the investigations of R.T. Berg and associates at the University of Alberta and Howard FREDEEN and associates at Agriculture and Agri-Food Canada. The results of long-term controlled selection studies in hogs by Stothart and Fredeen, in POULTRY by Robert Gowe

(Ottawa), and in beef cattle by Berg and scientists at Agriculture and Agri-Food Canada demonstrated the effectiveness of systematic selection, and were all in agreement with theoretical expectations. The Lacombe breed of hogs was the first livestock breed developed in Canada, by Stothart and Fredeen. This breed is popular in many countries around the world.

Statistical Procedures

During the past 2 decades, scientists at the University of Guelph (B.W. Kennedy, J.W. Wilton, L.R. Schaeffer), in cooperation with scientists at Agriculture and Agri-Food Canada, have made significant contributions in the development of modern statistical procedures to compare the genetic potentials of breeding animals (primarily males) by breed or by herd. These procedures use all relevant information available on an animal and its relatives in estimating its relative breeding value.

Performance Tests

Show-ring standards, once accepted as authoritative criteria for breeding merit, have been gradually replaced by performance tests that objectively measure the differences among promising breeding animals for traits such as growth rate and production of milk, eggs or wool. Performance test stations have been established across Canada to evaluate the individual performances of male animals, primarily hogs and beef cattle, under standard conditions for growth rate, feed efficiency and carcass merit. Livestock exhibitions and fairs now serve primarily to promote breeds and to sell breeding animals that have met performance test criteria.

Breeding Research

Animal breeding research in Canada proceeds on several fronts. Work at McGill has emphasized the genetics of congenital defects in MAMMALS, particularly humans. Guelph continues to contribute to the refinement of mathematical techniques for predicting the genetic potential of breeding animals. GENETIC ENGINEERING techniques (recombinant DNA) have been applied in the production of new vaccines for livestock diseases at the Veterinary Infectious Disease Organization at the University of Saskatchewan. Biotechnologists are trying to identify the genes with major physiological and biochemical effects in farm animals. Incorporation of "desired" genes into the genome without disrupting normal physiological processes will be the challenging task of biotechnologists in the future. Work at other research centres embraces both theoretical and applied aspects of POPULATION GENETICS in studies of mating systems, performance testing and the biological interpretation of genotypic and phenotypic variationAgriculture and Agri-Food Canada Online

An extensive information source about Canadian agricultural issues and related government programs. From Agriculture and Agri-Food Canada.

Canadian Food Inspection Agency

The Canadian Food Inspection Agency is mandated to safeguard Canada's food supply and the plants and animals upon which safe and high-quality food depends.

Holstein Canada

Check out the history, care, and breeding of the Holstein cow at the Holstein Canada website.

Canadian Council on Animal Care

The CCAC monitors the care and use of experimental animals in Canada. Their website offers detailed guidelines, data on animal use and related material. Sponsored by the Canadian Institutes of Health Research (CIHR), the Natural Sciences and Engineering Research Council (NSERC) and other agencies.

Fur Institute of Canada

This Fur Industry of Canada website focuses on such industry issues as animal welfare, humane animal capture devices and wildlife conservation. Also includes educational resources about fur bearing animals and sustainable use practices.

The Human-Animal Bond Association of Canada

HABAC promotes the benefits of human-animal relationships for senior citizens and others. Includes information about the Canadian Canine Good Citizen Test.

Glossary: Veterinary Medicine

A glossary of terms related to veterinary medicine. From Washington State University's College of Veterinary Medicine.

Centre for Prions and Protein Folding Diseases

This site offers brief descriptions of neurodegenerative disorders such as BSE (Bovine Spongiform Encephalopathy,"mad cow"), scrapie, Creutzfeldt-Jakob Disease (CJD), and Chronic Wasting Disease (CWD).

Canadian Cattlemen's Association

This site serves up the latest news about beef production in Canada. Features an extensive list of industry links about animal care, cattle identification systems, and more.

Meat Cuts Manual

Your illustrated guide to well dressed beef, poultry and other animal products. From the Canadian Food Inspection Agency.

Atlantic Cod Genomics and Broodstock Development

Information about research programs in Newfoundland and New Brunswick that focus on breeding cod stocks with good resistance to disease and stressors such as changes in water temperature, or effects of being handled.

Canadian Agri-Food Policy Institute

Check out this website for information and reports about current issues impacting on the productivity and competitiveness of Canada's agri-food sector.

FAWC Report on the Welfare Implications of Animal Breeding and Breeding Technologies in Commercial Agriculture

The Farm Animal Welfare Council (FAWC) was established in 1979. Its terms ofreference are to keep under review the welfare of farm animals on agricultural land, atmarket, in transit and at the place of slaughter; and to advise Great Britain's Rural AffairsMinisters of any legislative or other changes that may be necessary. The Council has thefreedom to consider any topic falling within this remit.

The aim of this report is to provide clear and practical advice to Government on theestablishment of an appropriate framework within which developments in animal breedingand breeding technologies, and the outcome of such processes, may be considered,monitored and, where necessary, regulated. FAWC, and a number of other bodies, have beenaddressing this issue for some time, the result of which for FAWC is this report. Thepublication of the Agriculture and Environment Biotechnology Commission's (AEBC) report on Animals and Biotechnology (2002) prompted Government to invite FAWC to establish ajoint working party with members of the Companion Animal

Welfare Council (CAWC), theAnimal Procedures Committee (APC) and a representative from the Department forInternational Development (DfID) to give further detailed consideration to the AEBC'srecommendations. Whilst FAWC intends to respond to the AEBC's recommendations in aseparate document, the welfare concerns and recommendations in this report are directlyrelevant to many areas covered by the AEBC report.

The commercial applications of new breeding technologies, as well as conventionalbreeding strategies, have the potential to influence animal welfare in a positive way. Forexample, in FAWC's *Report on the Welfare of Dairy Cattle (1997)* we recommended that,when commercially available, the sexing of sperm should be used to reduce the number ofunwanted male dairy calves, provided that the technique had not been shown to produceadverse effects.

Other potential 'positive' applications include breeding for longevity in dairycows, improved neonatal survival in pigs and breeding for anatomical characteristics to reduce the risk of fly strike in sheep. Breeding for disease resistance in a range of species isalso attracting increasing research interest.

On the other hand, inappropriate use of breeding technologies may create new problems, or exacerbate welfare problems that may already have arisen within conventionallivestock breeding. We have previously expressed concern that some of the serious welfareproblems in commercial agriculture are the outcome of a lack of balance in genetic selectionin conventional livestock breeding programmes. To highlight this, a summary of concerns and recommendations from previous FAWC reports is contained in Appendix A.

It is the impact of any breeding technology or strategy that is important to welfare, whether it is the quality of life of the offspring that is compromised, or whether it is the application of the technology itself that affects welfare.

Furthermore, where genotype associated welfare problems are recognised, FAWC believes there is no reason to separate commercial applications of new breeding technologies from conventional livestock breeding. Indeed, the boundaries between conventional breeding and biotechnology have become increasingly blurred, particularly as a result of developments such as marker assisted S selection which can allow faster genetic change in target traits than through conventional livestock breeding methods. Such developments should not necessarily be viewed as a threat to animal welfare. If they are applied to animal breeding in a responsible way, they have the potential to improve welfare.

Whilst FAWC is satisfied that the scientific development of breeding technologies developed within the UK is adequately controlled when under the Animals (Scientific Procedures) Act (A(SP)A), we believe that additional safeguards are required with regard to the suitability for their introduction into commercial agriculture. Additional safeguards are also required for the importation of new breeding technologies developed elsewhere, and for the importation of certain breeds of livestock, whether they are the product of new breeding technologies or the result of conventional breeding.

We are also of the opinion that there should be a proper assessment of welfare, not only for novel or existing technologies, whether imported or developed within the UK, but also for conventional breeding programmes. Given the concerns listed in appendix A, it may be seen that, in welfare terms, it is within the area of conventional breeding that some serious and extensive farm animal welfare problems are currently found in commercial agriculture.

This is a view supported by The Federation of Veterinarians of Europe (FVE) who considered these issues and in 1999 adopted a resolution urging, "member countries and the European Commission to consider the introduction of measures designed to safeguard the welfare of animals with respect to the risks inherent in selective breeding programmes, while preserving the unique characteristics and genetic advantages of European breeds".

In the production of this report we have used selected examples of welfare problems to illustrate the breadth of welfare issues associated with breeding and breeding technologies. More extensive reviews of techniques, recent developments in biotechnology, and the ethical and welfare issues associated with modern animal breeding may be found in reports by the Agriculture and Environment Biotechnology Commission (AEBC) *(Animals and Biotechnology, 2002)*, the Animal Procedures Committee *(Report on Biotechnology, 2001)*, the Royal Society *(The Use of Genetically Modified Animals, 2000)*, FAWC *(Report on the Implications of Cloning for the Welfare of Farmed Livestock, 1998)*, and the Banner Committee *(Report of the Committee to Consider the Ethical Implications of Emerging Technologies in the Breeding of Farm Animals, 1995)*.

FAWC'S Philosophy and Methods

Animals are kept for various purposes and in return their needs should be provided for. They are recognised as sentient beings in the Treaty of Amsterdam, thus FAWC considers that we have a moral obligation to each individual animal that we use. This obligation includes never causing certain serious harm to animals and, when deciding on our actions, endeavouring to balance any other harms against benefits to humans and/or other animals.

The achievement of high standards of animal welfare requires awareness of animal needs and both caring and careful efforts on the part of all that are involved in the supervision of farmed animals. General guidelines as to what those who use animals should provide in order to avoid suffering and other harms, are contained in the five freedoms:

- Freedom from hunger and thirst, by ready access to fresh water and a diet to maintain full health and vigour;
- Freedom from discomfort, by providing an appropriate environment including shelter and a comfortable resting area;
- Freedom from pain, injury and disease, by prevention or rapid diagnosis and treatment;
- Freedom to express normal behaviour, by providing sufficient space, proper facilities and company of the animal's own kind;
- Freedom from fear and distress, by ensuring conditions and treatment which avoid mental suffering.

When assessing any welfare problem, it is necessary to consider both the extent of poor welfare and its duration. Welfare assessment concerns individual animals. However, where there are indications of poor welfare, we consider that the more animals which are affected, the more serious is the problem. In order to offer useful advice about the welfare of farm animals, FAWC takes account of scientific knowledge and the practical experience of those involved in the agriculture industry. A broad-ranging approach, taking into account all relevant views and attempting to balance human benefit with a concern to ensure that the animal's interest remains to the fore, is used in the formulation of FAWC recommendations. Knowledge based on scientific studies of the welfare of animals is increasing rapidly. The term 'animal welfare' is employed frequently in scientific and legal documents and in public statements. In our view, welfare encompasses the animal's health and general physical condition, its mental state, its biological fitness and its ability to cope with any adverse effects of the environment in which it is kept.

FAWC first considered welfare issues associated with animal breeding in 1991 when we expressed concern that breeding had altered the shape of Belgian Blue cattle such that they were unable to reproduce naturally without an unacceptable degree of pain, and that a large proportion of calvings required a caesarean section for the calves to be delivered. We discussed the welfare implications of breeding technologies in 1993, when giving evidence to the Banner Committee, which had been established to consider and report on the ethical implications of emerging technologies in the breeding of farm animals. One of the main points FAWC made to the Committee was that we regarded the Animals (Scientific Procedures) Act 1986 (A(SP)A) as a sound Act, particularly because the principle of cost benefit analysis was applied to proposals, thereby assessing the extent of suffering to animals against the potential benefit to society and to other animals.

We thought that it had improved the attitude of researchers towards experimental animals by requiring them to demonstrate to the Home Office the integrity of the work and that there would be benefits from it. At the time, FAWC expressed concern that no comparable legislation existed to protect animals once they left the Act and entered commercial farming practice.

FAWC suggested that on release from A(SP)A, a technique needed to be in the public domain, needed to be monitored for 4-5 years and then reviewed in the light of comprehensive field studies. FAWC argued that the application of any new technology should be defined, not only by method, but also by consequence. We proposed to the Banner Committee that an independent body should be established which would deal with specific procedures by reviewing the welfare costs and benefits of their commercial application. We suggested that such a body need not be overly restrictive and that it would also help alleviate any public disquiet about such technologies by demonstrating that they had been properly considered.

We welcomed the publication in 1995 of the report by the Banner Committee. It provided an excellent overview of developments in biotechnology and produced a number of recommendations, which FAWC supported. The committee agreed with FAWC that there was a need for an independent body to rigorously address the ethical questions that future developments would pose. A key recommendation of the Banner Committee was that a standing committee for this purpose should be established. The Banner Committee also commented that although normal selective breeding fell outside of its remit, that it was not invariably neutral as regards animal welfare, and could result in "highly objectionable side effects". In 1998, FAWC published its *Report on the Implications of Cloning for the Welfare of Farmed Livestock*. We recommended that the general principles as prescribed by the Banner Committee should be adopted as a framework within which present and future uses of animals should be assessed. FAWC also made a series of recommendations specific to cloning technology. A central recommendation was that a National Standing Committee should be established to oversee developments in cloning technology.

More recently, FAWC gave evidence to the AEBC during their work towards their report *Animals and Biotechnology (2002)*. The report makes an important contribution to the debate on biotechnology, particularly since it has also sought to include public opinion in the development of its conclusions. FAWC has responded to the AEBC report in a separate document, and we support AEBC's main recommendations. In particular, the AEBC report recommended that "a new strategic advisory body should be set up to examine issues raised by the use of genetic biotechnology on farm animals in the context of its use on other animals and current livestock farming practices".

Regulations

Current animal welfare regulations and codes relevant to breeding and breeding Technologies

General Welfare Provisions

The overarching legislation covering animal welfare in England and Wales is the Protection of Animals Act 1911. In Scotland, the Protection of Animals Act (Scotland) 1912 applies. These Acts contain the general law relating to cruelty to animals. Broadly, it is an offence to cause any unnecessary suffering to any domestic or captive animal by anything that is done or omitted to be done. The Government is in the process of consolidating and modernising animal welfare legislation including the Protection of Animals Act 1911. FAWC understands that a major aim of this exercise is to promote, within the revised legislation, the ideas of good welfare standards and the promotion of proactive measures to ensure that animals do not suffer. It is expected that Scotland will introduce similar revisions to its animal welfare legislation. FAWC welcomes this revision of the legislation in all the administrations. The Agriculture (Miscellaneous Provisions) Act 1968, Part I, is the primary piece of legislation applicable specifically to farm animals. The Act makes it an offence to cause or permit livestock on agricultural land to suffer unnecessary pain or unnecessary distress. The Act empowers Ministers, subject to Parliamentary approval, to make mandatory regulations and to issue Codes of Recommendations for the welfare of livestock. They are also empowered to prohibit certain operations on animals.

Legislation made under the Agriculture (Miscellaneous Provisions) Act includes The Welfare of Farmed Animals (England) Regulations 2000 and the equivalent regulations in Scotland (2000) and Wales (2001) which contain the general conditions under which all farmed animals must be kept.

Many elements of these Regulations implement European Directives on farm animal welfare. For example, an amendment to the Regulations was introduced in 2002 to include the requirements of an updated European Directive applicable to laying hens. Most farm animal welfare legislation is now being set by the European Union for implementation by all Member States. The Department for Environment, Food and Rural Affairs (Defra) produces Codes of Recommendations for all those species specified under the Regulations. These Welfare Codes contain guidance to encourage those who care for farm animals to adopt acceptable standards of husbandry. They also set out the important legal requirements.

EU Legislation on Farm Animal Breeding Procedures

Specific legislation on farm animal breeding procedures is now in force as a result of European Directive 98/58/EC concerning the protection of animals for farming purposes. This is implemented in The Welfare of Farmed Animals (England) Regulations 2000, and the equivalent Regulations for the devolved administrations, which state that: "natural or artificial breeding procedures which cause, or are likely to cause, suffering or injury to any of the animals concerned shall not be practised", and that: "no animal shall be kept for farming purposes unless it can reasonably be expected, on the basis of their genotype or phenotype, that they can be kept without detrimental effect on their health and welfare."

Domestic Legislation Specific to the Development of Novel Breeding Technologies

The process of developing any new breeding technology within the UK is covered by The Animals (Scientific Procedures) Act 1986 (A(SP)A), which is administered by the Home Office in England, Scotland and Wales. This Act is for the protection of animals used for experimental or other scientific purposes. Animals covered under the Act are all non-human vertebrates, including larval or embryonic forms that have reached a certain stage in development. The Act also covers one invertebrate species (Octopus vulgaris). The Act covers work on Genetically Modified (GM) animals and also cloned animals. Under the Act, any scientific procedure which may cause pain, suffering, distress or lasting harm is judged to be a regulated procedure. In the case of GM animals, both the initial production as well as any subsequent breeding of the animal is considered a regulated procedure. The legislation applies until the death of the animal and applies to its offspring, should there be any, until their death. Second generation offspring from these animals may be considered for discharge from the Act following submission of acceptable welfare records covering the full natural lifespan of the animal. FAWC is satisfied that the regulations are strictly enforced and that animals would not be released if there was any evidence that welfare might be compromised. At the time of writing this report FAWC is not aware that any applications have been received for discharge of GM animals from A(SP)A. There is additional EU legislation covering the development, placing on the market, traceability and labelling and transboundary movements of all Genetically Modified Organisms (GMO's) which adopts a step by step process for their assessment. Initially, genetic modification is carried out in containment. This is governed by EC Directive 90/219 (as amended by Directive 98/81/EC) and is closely controlled by the Health and Safety Executive. This Directive is implemented in the UK by way of the Genetically Modified Organisms (Contained Use) Regulations 2000, the GMO (Risk Assessment) (Records and Exemptions) Regulations 1996 (as amended) and the Environmental Protection Act 1990.

The deliberate release and placing on the market of GMOs, including animals, is governed by European Directive 2001/18 EC. In Great Britain this has been implemented by way of Part VI of the Environmental Protection Act 1990 and, in England, the Genetically Modified Organisms (Deliberate Release) Regulations 2002 (equivalent regulations have been implemented in the devolved administrations). Decisions on whether or not to allow a release are based on a detailed assessment of any risks that may be posed by the GMO to human health or the environment on a case-by–case basis. This would also include assessment of any risks to animal health, although not specifically welfare. Directive 2001/18 also includes requirements for labelling of GM products. This Directive has been amended by two new EU regulations on traceability and labelling (1830/2003) of GMOs and GM food and feed (1829/2003) which came into force in April of this year. Any application in the future to assess a GM animal for food use would therefore be assessed by the European Food Safety Authority under the new food and feed legislation.

Other welfare initiatives relevant to animal breeding

Within the UK there are examples of initiatives by animal breeders as well as veterinary bodies to address welfare concerns associated with breeding and breeding technologies. For example, the UK dairy industry has developed a nationally available selection index (£Production Lifespan Index (£PLI)) that incorporates longevity, as an inclusive measure of cow health, in addition to production traits. Plans are in place to expand the £PLI to include additional health traits, for example, lameness, that will increase opportunities for dairy farmers to select bulls for both health and production. FAWC is encouraged by these developments. The Sheep Veterinary Society and the British Cattle Veterinary Association (BCVA), amongst others, have both produced guidelines on advanced breeding technologies which recognise welfare concerns to reduce certain breeding associated problems. Such initiatives and Codes of Practice are to be welcomed, and FAWC would encourage all sectors to follow their example. However, their impact on welfare will be determined by the extent to which they are adopted and applied in any given livestock sector.

The Royal College of Veterinary Surgeons (RCVS) advises on artificial breeding techniques, including embryo collection and transfer, in its Guide to Professional Conduct (2004). The advice is based upon the Bovine Embryo (Collection, Production and Transfer) Regulations 1995, but through its professional guidance, the RCVS extends the principles to other species and techniques used in advanced breeding technology. It is stressed that, at all stages in such procedures, the welfare of animals should be paramount. Nevertheless, the RCVS has no mechanism to routinely monitor compliance with this advice.

Furthermore, this advice is only applicable in techniques where veterinarians are directly involved or are responsible for supervision. The Government and their devolved counterparts, the Scottish Executive (SE) and the Welsh Assembly Government (WAG) have recently published their Animal Health and Welfare Strategy for Great Britain. One of the aims of the strategy is to encourage industry to develop animal health and welfare plans, something which has the potential to impact on animal breeding and breeding technologies significantly. FAWC welcomes this joint initiative and hopes that both health and welfare will be given equal emphasis.

European Initiatives

In response to growing public concern about farm animal breeding and reproduction, the Sustainable European Farm Animal Breeding And Reproduction (SEFABAR) project was initiated in 2000 by the Farm Animal Industrial Platform (FAIP). It was an EU funded Thematic Network of representatives from all sectors of the livestock industry, breeding scientists and economists, brought together in a series of workshops over a three year period. During this time, the remit of SEFABAR was to discuss the future sustainability of livestock breeding within Europe, including a consideration of future European and world markets. Animal, human health and environmental considerations also formed important parts of the discussions. One of the outcomes of the workshops is the agreement by breeding organisations represented within SEFABAR to develop Codes of Practice for farm animal breeding. These codes are now being developed under a new 18-month FAIP co-ordinated project, Code of Good Practice for European Farm Animal Breeding and Reproduction (CODE-EFABAR). A draft of these Codes is expected in September 2004. If welfare is given a high priority within these proposed Codes and European breeding organisations agree to operate within them, they have the potential to raise the prominence of animal welfare as a key issue in changing breeding strategies. However, it must be recognised that many breed organisations operate within world markets and this may constrain the degree to which such Codes may address welfare concerns, particularly those which, in order to enhance welfare, might constrain the ability to achieve the gains that commercial sustainability usually requires.

FAWC welcomes the SEFABAR and CODE-EFABAR initiatives and their outcomes to date, in particular, the fact that SEFABAR also considered ethical standards and recognised the importance in animal breeding of selecting for welfare enhancing traits. However, it remains the case that such initiatives are industry led and thus, in that sense, do not satisfy the requirement for independence that was called for in the Banner Committee, AEBC and FAWC cloning reports in their recommendations for a group to consider farm animal breeding issues.

Gaps in Current Regulations

Concerns About General Welfare Legislation

FAWC recognises the value of the EU legislative requirement specific to animal breeding (paragraph 27) but is concerned about how effectively it is enforced. For example, we are not aware of any cases where it has been used successfully to restrict any breeding procedure. Examples of genotype associated welfare problems in commercial agriculture, such as those documented in the modern dairy cow or broiler chicken, demonstrate the obvious difficulties in defining what is unacceptable in terms of animal welfare. It is also clear that when problems are recognised in species in widespread commercial use, there may often be no easy solution to rectify them, particularly when they have arisen as a result of past breeding strategies or changes in husbandry and management. Effective advice, and possibly legislative control, is needed to define acceptable and realistic breeding goals if such welfare problems are to be addressed.

FAWC has also sought to determine how those sections of European Directive 98/58/EC concerning animal breeding are interpreted and implemented in other parts of Europe. However, we have found no detailed regulatory framework in any Member State which addresses fully the particular problems associated with the breeding of farm livestock for commercial purposes.

Member States such as Italy have taken a similar approach to the UK in that the wording of the European Directive has been incorporated into national legislation. Denmark and Sweden have introduced legislation which allows the possibility of future controls. For example, the Danish Act on the Protection of Animals 1991 states that the Minister of Justice may lay down rules prohibiting the release of bred animals which have difficulties living in nature. A further provision gives the Minister of Justice the power to lay down more detailed rules on biotechnology, including a prohibition on the use of such methods on animals kept for farming purposes. German animal welfare law attempts to define more precisely the nature of problems associated with breeding which are considered unacceptable. It is prohibited to breed vertebrates or to change them through biotechnology or genetic engineering if it is expected that the offspring are lacking parts of the body or organs for species specific use or they are unfit or deformed thereby causing pain, suffering or harm. The German legislation specifically mentions behavioural and other welfare problems and prohibits the production of vertebrates where it is expected that behavioural abnormalities will occur resulting in suffering or increased aggressiveness. The law also prohibits breeding vertebrates if their keeping is only possible under conditions causing them pain, avoidable suffering or harm. We conclude that the lack of an adequate framework within the UK for the detailed consideration of how European Directive 98/58/EC may be interpreted and enforced is a significant gap in current welfare controls.

Concerns About Ethical issues associated with animal breeding

The AEBC report expressed concern that there is a potential gap in the existing welfare legislation in relation to "the generation of what might be judged intrinsically objectionable changes to animals" even in the absence of clear animal welfare, animal or human health, or environmental concerns, as applicable to both GM and conventional farm animals. The report stated that such "intrinsically objectionable changes" would include insentient animals or animals with their physical characteristics, or normal patterns of behaviour, radically and unacceptably altered.

A key recommendation from AEBC was that provision within legislation will be needed to protect animals from such developments. The report of the Banner Committee illustrated the above point with a hypothetical example of a breeding strategy, the aim of which was to produce pigs of reduced sentience and disinclined to engage in activity normal to them. The report stated, "even if this has no welfare implications (if welfare is understood narrowly as relating to an animal's happiness), so that by any available measure such pigs are as content as any other pigs, still we would maintain that the proposed modification is morally objectionable in treating the animals as raw materials upon which our ends and purposes can be imposed regardless of the ends and purposes which are natural to them. The fact that the project promises an increase in profit, or any other desirable consequence, does not, and cannot, wipe out the intrinsically objectionable character of such an action". FAWC agrees with both the Banner Committee and with the recommendation of AEBC, that serious consideration must be given to such complex ethical issues.

Concerns about the Importation of Animals

FAWC is satisfied that the existing controls under A(SP)A, as administered by the Home Office, are adequate to protect the welfare of animals in the laboratory and to control the development of new technologies within the laboratory environment. There is general acceptance, however, that there are potential gaps in control which may allow techniques and animals to be released into commercial agricultural practice unchecked. During our consultation process we were told that no satisfactory regulations are in place to control the importation of GM animals or embryos for commercial agriculture. The Home Office authorises the acquisition and use of a GM animal imported into the UK for scientific purposes (therefore involving A(SP)A)) although the actual importation is controlled by Defra. The importation of GM animals and embryos for agricultural purposes are governed by Directive 2001/18, Regulation 1829/2003 and 1830/2003. These regulations require that all GMOs must have relevant authorisation and be appropriately labelled at all stages of placing on the market. In addition, transboundary movements of GM animals from one country to another are also covered by the relevant requirements for safe transfer, handling and use of the Cartagena Protocol on Biosafety. Regulation (EC) No 1946/2003 of the European Parliament and of the Council on transboundary movements of GMOs establishes a common system of notification and information for transboundary movements of GMOs.

Despite the controls listed above (and in paragraphs 31 & 32), since the GM status of every animal or consignment of animals entering the country is not currently declared on the paperwork that typically accompanies those animals, although illegal, it would be possible for a GM animal to enter unchecked. That is, since a GM animal does not differ in outward appearance from a non-GM animal, the possibility still exisits that the illegal import of GM animals could realistically occur. In addition, even within the law, it is of concern that no specific assessment of an imported GM animal's welfare is required as part of the importation procedure. Given the above, FAWC is concerned that there are potential gaps in the current regulations which would allow GM animals to enter the UK without official knowledge and/or due consideration of their welfare, and we should like the Government to consider how these potential loopholes may be closed.

Recommendation

FAWC recommends that the Government consider methods to close potential loopholes that would allow GM or cloned animals, their gametes or embryos, to enter UK commercial agriculture uncontrolled.

Concerns About the Importation of Novel Breeding Technologies

A problem in the case of novel technologies is that many are developed from commercial sources, often overseas, and are therefore not initially covered by A(SP)A. Technologies can be introduced from overseas by veterinary surgeons as part of "recognised veterinary practice". These could become established within livestock farming before there had been any proper evaluation of welfare implications.

In the Banner report, ovum pickup was cited as a particular area of concern. This potential problem is also well illustrated by juvenile in vitro embryo transfer (JIVET), a technique currently used commercially in Australia. JIVET is the mechanism through which follicle growth in juvenile animals (calves of 8-10 weeks old and sheep and goats of 6-8 weeks old) can be stimulated, offering the potential to substantially reduce generation intervals and produce multiple progeny. Practically, the technique requires hormone treatment of prepubertal animals, followed by oocyte recovery under general anaesthesia and via laparoscopy. Although this procedure, which presents clear ethical questions and may carry potential welfare problems, is not currently used in UK commercial agriculture, the possibility that this may become the case, as in Australia, is real. Methods of detecting such imports, perhaps through liaison with veterinary practices and organisations, breeding and agricultural representative organisations, and Government departments will be important. In addition, the continued monitoring of imported techniques for an extended period following their introduction is important to ensure that welfare problems which may exist, but which may not be immediately obvious at the time of import, are detected further down the line.

Concerns About Domestically Developed Dreeding Technologies

Even for technologies developed within the UK, once they are outside the protection of A(SP)A, any animal that is subjected to or is the product of new technology is protected only under the general welfare legislation. For example, concern has been expressed to FAWC about the consequences of initiatives to promote the incidence of twin calves in the beef industry through the implantation of multiple embryos. Whilst the technology required to achieve this may not be, in itself, a welfare concern, we are aware that problems, such as poor calf survival and disease have arisen in some commercial agricultural systems. Additional welfare problems may be associated with the implementation of breeding technologies already in existence. For example, there are no rules to govern the number of embryos which may be implanted into sheep or cattle, or the number of times such a procedure may be performed. Surveillance (both passive and active) by bodies such as the State Veterinary Service, Veterinary Laboratories Agency (VLA) and the Meat Hygiene Service (MHS) may be effective in discovering welfare problems such as these but it is recognised that this is a reactive, rather than a proactive approach. Furthermore, current resources available to such agencies will limit the level of detection achieved.

The limitation of resources also makes it impossible to inspect regularly and effectively a sufficient proportion of agricultural holdings. Compared with the situation in research establishments, where every licensed laboratory is subject to at least an annual visit by the Home Office Inspectorate without prior notice, farms might go for many years without any inspection. This is a potential problem for both the detection of novel techniques as well as for the monitoring of welfare problems that might arise in commercial practice. The Veterinary Surveillance and Animal Health and Welfare Strategies currently being developed jointly by Defra, SE and WAG represent a good opportunity to address these problems and improve the effectiveness of the surveillance system, as long as they are adequately resourced.

It is essential that targeted surveillance is made of farms where new technologies, developed under laboratory conditions, but recently released into commercial practice, have been implemented. There is a strong argument for a period of commercial trials before novel techniques may be available for general use. This would provide a bridge between the controlled conditions of the laboratory and general farm use. FAWC raised this issue in 1993 when giving evidence to the Banner Committee. We proposed that new techniques needed to be monitored in comprehensive controlled field studies for 4-5 years, and then reviewed in the light of the evidence gathered.

Concerns About the Development of Clinical Veterinary Techniques

Veterinary medicines are strictly controlled and cannot be used commercially until evidence of safety, quality and efficacy, as well as any identifiable welfare consequences have been determined. The same is not the case for clinical practices. These can be developed by veterinary clinicians and, in the absence of a formal mechanism to review their efficacy and welfare impact, may be introduced into commercial practice unchecked.

Welfare Considerations

Welfare Consequences of Animal Breeding

60. Since 1992, all FAWC reports on the welfare of different species of livestock have highlighted welfare concerns associated directly with animal breeding strategies. However, compared to many other issues which FAWC has addressed, it has been far from straightforward to offer useful advice or to make recommendations as to how such problems may be resolved. The following examples illustrate the range of issues about which we have previously expressed concern and which still need to be addressed.

In 1997 *(Report on the Welfare of Dairy Cattle)* FAWC expressed concern that in the modern dairy cow, selection for increased milk yield had compromised welfare, reflected by an increased susceptibility to lameness and mastitis and a reduction in fertility. Subsequent research has supported this conclusion. We recommended that breeding companies should devote their efforts primarily to selection for health traits so as to reduce levels of lameness, mastitis, and infertility and that selection for milk yield should follow only once these health issues have been addressed. The report also made the general recommendation that breeding programmes worldwide should have as a major objective the need for good welfare. Selection criteria used should also not lead to the production of animals that require above average levels of management to prevent welfare problems thus providing a degree of safeguard where management is less than optimal

The FAWC *Report on the Welfare of Dairy Cattle* also made a number of recommendations specific to welfare concerns associated with breeding technologies such as ovum pick-up, repeated epidural injections for oocyte collection, the effects of repeated administration of superovulatory drugs, and the problems regarding oversized calves, and hence calving difficulties, resulting from in vitro fertilised embryos.

FAWC recognises that multi-trait breeding uses complex mathematics to arrive at an economic optimum for the selection pressure applied to individual traits, and therefore that the above recommendations may lead to the selection of animals that are not performing at the commercial optimum, thus creating a difficulty for breeding companies. In addition, it is recognised that since current breeding indices are at best holding steady on health traits, it may be difficult to reverse problems which have already emerged.

The example of the dairy cow demonstrates the need for a broad strategic approach to addressing welfare problems associated with genotype. Such an approach must, of necessity, involve the co-operation of breed companies, farmers, geneticists, veterinary and other advisory organisations. There is an argument that if real welfare improvements are to be made, there is a need for some level of independent advice, and possibly regulation, on the genotypes that are being promoted within commercial agriculture.

Welfare problems associated with conventional breeding methods are also demonstrated in the modern broiler chicken where there is evidence to link past selection for fast growth with associated leg and cardio-pulmonary problems. The FAWC *Report on the Welfare of Broiler Chickens (1992)* raised particular concerns about the level of leg problems and proposed four principle methods of reducing the incidence, including the increased selection of breeding stock for strong and well-formed legs. Recognition of such problems has encouraged broiler breeding companies to modify selection programmes. However, there is a need for assurance that these changes have had positive effects on animal welfare.

FAWC also commented on the selection of broiler breeding stock in its *Report on the Welfare of Broiler Breeders (1998)*. We emphasised the importance of ensuring that factors such as cardio-vascular health, foot and leg health, social behaviour and resistance to disease were given high priority in selection procedures. We also expressed concern at the problem of hunger in broiler breeders and recognised that it was likely to get worse if selection for fast growth continued. We made the specific recommendation that the objectives of the breeding companies in the future development of strains of broilers should include welfare improvement, in particular the avoidance of problems of prolonged hunger in broiler breeders. The resolution adopted in 1999 by the FVE summarised their concerns associated with animal breeding. They stated that "Selective breeding programmes may cause animal welfare problems. It may become difficult or impossible for natural copulation or parturition to occur; offspring produced by selective breeding for certain specific characteristics may be unable to express their natural behaviour; or they may be predisposed to hereditary, congenital, metabolic or infectious disease, disability or early death. The introduction of such selective breeding programmes may make it impossible for the breed to be maintained by natural means".

On the subject of breeding technologies, the FVE stated that, "the use of new and emerging technologies in artificial breeding, such as ovum and embryo transplantation and genetic manipulation, may also be a source of concern, and it is likely that some future advances in science will also have animal welfare implications. The technique used may carry inherent welfare risks for the animal (e.g. the particular method by which semen or ova are obtained); the intended outcome of the procedure may be intrinsically objectionable (e.g. the development of animals with unnatural physical or behavioural characteristics); and offspring may be produced with welfare disadvantages such as those mentioned above".

A recent report published by the Department of Trade and Industry (DTI), 'Genetics and Genomics of Sheep and Cattle in Australia and New Zealand' effectively highlights the "technological crossroads" that animal breeding has reached. The report emphasises that, "new opportunities are opening up that are likely to transform the way breeders improve their stock", for example, growing commercial interest in the potential of marker assisted selection looks set to accelerate the rate of genetic change to livestock by conventional selection methods. FAWC recognises that the application of gene-mapping to selective breeding programmes may be used to rectify recognised welfare problems, for example, by selecting for specific health traits such as improved leg health in broilers. We are concerned, however, that with the considerable commercial competition between breed companies, the primary focus of attention will be for production-related traits. In the case of the dairy cow this might be for higher milk volume and changes in constituents, and for the broiler chicken, faster growth rate, improved feed conversion ratio, or greater breast muscle mass.

We are aware of research groups using marker-assisted selection for animals with greater levels of disease resistance, for example, salmonella resistance in poultry and parasite resistance in sheep. The Dti report also states that in Australia and New Zealand, "there [is] considerable interest from a number of groups to identify and exploit genetic variation among livestock for disease resistance". Whilst this will have obvious welfare benefits, it is important that the development of such strains is not used to disguise welfare threatening conditions which would otherwise produce disease and does not discourage the development of higher standards of stockmanship and provision of a good quality environment.

Genotype and Environment Interactions

The selected examples of welfare problems described in the previous section are those where narrow breeding objectives, or novel breeding technologies have had adverse consequences for animal health and welfare. However, breeding related welfare problems cannot be viewed in isolation since most are inextricably linked with the environment in which animals are kept. Of fundamental importance is the quality of management of any animal throughout its life, but there are many other aspects of the environment which, if inappropriate for a particular genotype, may have consequences for welfare which are just as serious as poor management. For example, welfare problems may arise where a particular breed of animal is poorly suited to the environment in which it is reared. FAWC has raised concerns about this in a number of recent reports.

In the *Report on the Welfare of Sheep (1994)* we expressed our concern about the potential problems associated with changes in breed structure in response to the commercial demand for different carcase conformation characteristics. We recommended that if any change in breed or breed type is contemplated in challenging extensive conditions, replacement must only be with one that is sufficiently well adapted to the environment. We also recommended that within breed selection programmes, monitoring is carried out for problems associated with selection for greater muscularity.

We made a similar recommendation in our *Report on the Welfare of Pigs Kept Outdoors (1996)* where we stated that, breeding companies, and those responsible for the selection of breeding stock to be kept in outdoor enterprises, should ensure that only those strains of pig with the genetic potential to thrive in the conditions are used. In the report the importance of temperament was also raised and we recommended that when selecting pigs, attention should be paid to the need for good temperament and mothering ability.

We hold the general view that the welfare of some breeds of high performance potential may be adversely affected when kept in more extensive or organic environments. The increasing demand for organically produced food has encouraged greater interest in this aspect of animal welfare with some research directed towards the suitability of breed types for organic systems.

For example, a recent study has examined the suitability of two commercial broiler strains, one fast-and one slow-growing, in a free range system. Both strains became very heavy at the minimum age of slaughter specified by organic requirements with the fastgrowing strain having the poorest feed conversion ratio. This, in addition to poor mobility, as reflected in low usage of the outdoor area, and the presence of deep pectoral myopathies led the authors to suggest that the fast-growing strain was particularly unsuitable for free range production. Given that organic standards require chickens to be slaughtered at a greater age than is now the normal age for standard broiler production, it is likely that exposing certain commercial broiler strains to such systems would be a welfare concern.

We have discussed with the Soil Association the potential problems which might arise from unsuitable genotypes in organic farming systems, and are encouraged that they wish to address this matter in future versions of organic farm certification standards. FAWC made a general comment about this matter in the *Interim Report on the Animal Welfare Implications of Farm Assurance Schemes (2001)* where we recommended that consideration be given, in particular, to the incorporation of scheme standards which relate to the breeding and rearing of animals for specific production systems.

We believe breed companies should take greater responsibility for this matter and we are concerned that environments in which breeds are developed are often far removed from those in which animals are subsequently reared when in commercial production. With breed companies increasingly operating in global markets, the potential welfare problems resulting from a mismatch between genotype and environment are likely to increase. For example, during consultations for the preparation of this report, concern was expressed about the export of high producing genotypes to situations where, for example, appropriate feeds to meet metabolic demands may not be available. Work on sheep breeding programmes at the Scottish Agriculture College (SAC) has incorporated traits that are important for being a 'good' ewe alongside those important for being a 'good' lamb, in addition to key sustainability traits such as lamb survival and ewe longevity. The new breeding indices developed are intended to improve flock efficiency without detriment to mothering ability and survival characteristics in extensive farming situations. Whilst the researchers recognise that their approach to animal breeding will have particular appeal to organic producers, they also recognise the value of the work to the wider UK hill sheep industry where, in order to offset falling incomes, there has been a trend to increase the number of sheep per shepherd and also a reduction in the amount of veterinary medicines used for disease prevention and treatment.

The standard of management is an aspect of the environment in which an animal is kept and we recognise that, with high levels of management, many of the genotypes of higher production potential can often be reared without major welfare concerns. However, FAWC is concerned about the necessity of the high levels of skill required by those persons responsible for some genotypes given the known variation in standards of management across farms.

FAWC recognises and welcomes the attempts made by industry to improve the management provided to emerging genotypes, and encourages the maintenance of research and training programmes for the development of these. However, given the potential importance of interactions between specific genotypes and environments on welfare, FAWC would suggest that, in addition, greater consideration of genotype and environment interactions in future breeding programmes is made.

Recommendation

FAWC recommends that industry, possibly with Government support, should sponsor research and training programmes for the development of husbandry systems to support the demands of new genotypes in relation to their production system.

Welfare Surveillance

Since breeding strategies, either by conventional breeding or using novel technologies can have such major influences on animal health and welfare, it is essential to have accurate information on the extent to which any trait which influences welfare is improving or getting worse, in addition to the respective impacts of genetic and environmental factors.

Breeding companies test the performance of new genetic strains under highly controlled conditions with very high standards of management, sometimes under the additional control of A(SP)A. It is on release to the commercial sector, when breed company management guidelines are sometimes ignored, standards of husbandry might be lower, or livestock are reared in less than optimal environments, that welfare problems often become apparent.

FAWC has discussed with a wide range of interested parties the difficulty of identifying breeding associated welfare problems which may arise. From these discussions it was evident that the lack of accurate on-farm information relating to health and welfare is a serious constraint on improving livestock welfare through breeding programmes.

A recent poultry industry survey of lameness in UK broiler chickens demonstrates the possibility of both cooperation and welfare surveillance within this sector of the livestock industry. However, it also demonstrates the potential limited value of such surveillance if those data collected are too restricted. For example, in this instance it remains unclear from the results whether the changes in leg health are the outcome of breeding programmes, or the result of general improvements in standards of husbandry and other environmental factors. This highlights the general situation that, since management and husbandry methods are constantly adjusting to the greater demands of modern genotypes, it may be difficult to determine whether improvements in welfare on the farm are the result of environmental changes rather than a direct effect of changing breeding goals.

Reliable monitoring of a range of welfare measures, together with carefully selected production, management and other environmental factors, is required to demonstrate the relative genetic and environmental contributions to altered welfare. If robust data were collected for all species over an extended period, a dissociation between those factors could be more easily achieved and would be of value to many, including breed companies, researchers and Government departments. For example, robust surveillance data collected prior to and following the withdrawal of meat and bone meal or in-feed antibiotics for poultry production may have aided industry in their evaluation of any reported welfare consequences of such changes. The importance of welfare surveillance to animal breeding strategies has been demonstrated in Scandinavia where, for the last 20 years, integrated databases and comprehensive recording schemes have been developed for both cattle and pig breeding. In the 1970s Scandinavia developed a philosophy that breeding objectives should include health and production traits rather than just production goals. It was recognised that an essential prerequisite for the efficient operation of such breeding objectives was the accurate recording of health, reproduction and production traits. Integrated databases, initially between the milk-recording scheme and the artificial insemination (AI) service, were developed and subsequently expanded to include health traits. For example, in all Scandinavian countries, veterinary reports on clinical treatments are now incorporated into the databases. The result is that Scandinavian countries have adopted Total Merit Indices (TMI) in selection programmes. Not only has such an approach improved animal health, as demonstrated for example, by a steady decline in mastitis levels in dairy cattle, but the total economic gain from selection for a TMI in dairy cattle has been shown to be 10-25% superior to single trait selection, despite a reduced gain in milk production levels.

The Scandinavian model has shown the importance of integrated databases and comprehensive recording schemes. The information obtained has provided effective management tools at farm level with economic benefits; it has produced valuable information for research and development at a national level; and it has provided a vehicle for the application of research findings into commercial practice. The cooperative structure of farming in Scandinavia has facilitated the development of such recording systems and FAWC recognises that the different structure of the farming industry within the UK may make the collection and sharing of information more difficult. However, FAWC is encouraged that such difficulties are being addressed and that the breed companies and breed societies have been directly involved in the development and implementation of this.

Within the UK, the State Veterinary Service, Local Authorities, and other bodies, monitor compliance with Welfare Codes and legislation. However, welfare surveillance at this level is mainly designed to expose the minority of producers where standards are poor or unacceptable, rather than to provide information about more subtle changes in the prevalence of welfare problems. Simple, workable recording systems for all species covering a range of measures agreed by all interested parties are required to produce robust data with the ability of exposing these subtleties.

In some livestock sectors, much of the desired information is already being gathered by, for example, breeding companies. This should be utilised and supported by additional monitoring and surveillance where necessary. However, it is essential that although data may be obtained from a range of sources, their analysis must be carried out by a body which is considered by all to be independent. In addition, where data is not of a confidential nature this should be made available for further analysis by interested parties.

In summary, it is FAWC's view that there is an urgent need to develop for the UK, on-farm welfare surveillance systems, capable of providing reliable, robust information on the prevalence of a range of health and welfare traits for different species of livestock. The information obtained from such surveillance systems would be of value to, and must be available to, farmers, breed companies, veterinarians and researchers. Whilst we recognise that production data will be one element of the data collected, the aim of the surveillance must always be to monitor and improve welfare. The recently published Animal Health and Welfare, and Veterinary Surveillance Strategies are well timed to ensure that this balance is achieved.

Genetic Modification

This report uses the same definition of genetic modification as that used by The Royal Society in their report, *'The use of genetically modified animals' (2001).* The term 'GM animal' refers to animals modified either via transgenesis (when individual genes from the same or a different species are inserted into another individual) or by the targeting of specific changes in individual genes or chromosomes within a single species.

The Royal Society Report (2001) summarised the technical barriers that had to be overcome before the production of GM livestock for food production would be viable, notwithstanding its acceptability to the public. These include; the low efficiency of genetic modification of the genome for pigs, sheep and cattle; the high levels of embryonic loss; the incomplete knowledge of the genome for most of the major farmed species; and the fact that potentially desirable traits such as disease resistance and improved production are polygenic and require the alteration and co-ordinated expression of several genes. The report also noted that funding agencies were not supporting GM livestock projects to a high level since investment returns were considered to be low. They concluded that the commercial development of GM animals as a source of food was unlikely to be progressed unless the regulatory, ethical, economic and environmental issues, as well as public concern can be addressed.

The DTI Global Watch Mission report, Genetics and Genomics of Sheep and Cattle in Australia and New Zealand, published in 2003, highlights how this field has developed and expanded since 2001. In contrast to the predictions of the Royal Society, the DTI report states that "over the next 20 years, the number of genes with identified variation relevant to livestock is set to increase dramatically as the knowledge of genes and their functions increases at a phenomenal rate". A key recommendation from this report is that "the UK (industry and Government) should consider significant new investment in molecular genetics of cattle and sheep".

The extent to which genetic modification will become incorporated into future livestock breeding strategies may well be determined, not by scientific developments, but by public acceptability of the technology. Opposition to GM crops by consumers, retailers and environmentalists continues to influence the commercial application of GM technology in the plant sector, and there is no reason to believe that a similar level of opposition would not develop if the technology became incorporated into livestock breeding. Given the above, and also the rapid pace of developments in this area, FAWC recognises the need to remain informed regarding this issue.

Cloning for Commercial Purposes

Although cloning may be used in conjunction with genetic modification technology, it is fundamentally different in that a clone is an organism or cell derived from a single ancestor by asexual means. It was the production in 1997 of a cloned sheep (Dolly) from an adult cell that resulted in considerable public debate on the implications of cloning, particularly the wider ethical issues.

In 1998 FAWC produced a *Report on the Implications of Cloning for the Welfare of Farmed Livestock*, which considered the welfare implications of the techniques involved and the regulatory controls which might be necessary. FAWC considered both the ethical and welfare issues associated with cloning and made a number of important recommendations. One overriding recommendation was that, until the problems of oversized offspring, embryonic and foetal losses and birth abnormalities, and the possibility of problems associated with aged DNA, have been satisfactorily resolved, there should be a moratorium on the use of cloning by nuclear transfer in commercial agricultural practice.

We also recommended that a National Standing Committee should be established to oversee the developments of cloning technology. It was stated that the Committee should review outputs of research aimed at tackling the welfare problems identified in FAWC's Cloning Report (and any other problems which may emerge); it should determine the time when it may be appropriate to introduce cloning into commercial agricultural practice; and it should ensure that the controls put in place at that time are both adequate and effectively implemented. The report also recommended that the National Standing Committee should play a role in both promoting public awareness of the facts and issues concerning cloning and related technologies, and conveying public concerns to Government and Scientists.

The problems associated with cloning identified in FAWC's 1998 report still remain. In all species the efficiency of the technology is still very low: for example in cattle, which is the most studied species, on average only 3% of the transferred cloned embryos develop into viable calves. There are a number of welfare problems associated with nuclear cloning. For example, clones tend to have higher birth weights and may have a greater propensity in later life for respiratory problems and immune system deficiencies compared with normal animals. In addition, placental and foetal abnormalities that can lead to death of the clone at various stages of development are common.

In response to FAWC's Cloning Report, Defra supported many of FAWC's recommendations. However, because commercial applications of cloning were still seen as not of immediate concern, it did not see a need for a moratorium on the commercial use of cloning by nuclear transfer, as proposed by FAWC. However, Defra did not rule out the establishment of a National Standing Committee to oversee the development of cloning technology.

It is difficult to predict the extent to which cloning will become incorporated into food animal production in the future. Research has suggested that because of the current technical and welfare problems, there will be few practical applications of cloning in commercial agriculture in the foreseeable future. However, representatives of commercial breeding companies developing cloning for commercial applications see many potential benefits and have predicted that cloning will become a routine part of livestock breeding within 20 years. They suggest that cloning will serve a number of purposes such as the commercial development of disease resistant animals, improved feed conversion, greater muscle mass, and the production of meat of more consistent quality. Breed companies also see an application of cloning to evaluate the performance of animals of the same genetic make-up under different management systems and also in preserving the genome of both premium and rare breeds of animals.

At the present time, within the UK, all cloning work including any work on possible commercial applications, is confined to research establishments and is done under the protection of A(SP)A. However, in the light of the predictions made by some commercial breeding companies involved in cloning work, it is essential that FAWC keep a close watching brief on developments in this field.

Ethical Considerations

FAWC addressed the subject of ethical aspects of biotechnology in the *Report on the Implications of Cloning for the Welfare of Farmed Livestock (1998)*. We referred to the general principles of the Banner Committee and we adopted its ethical framework in that a procedure may be considered intrinsically objectionable for any one of the following reasons:

a) It results in very severe or lasting pain on the animals concerned;

b) It involves an unacceptable violation of the integrity of an animal;

c) It is associated with the mixing of kinds of animals to an extent which is unacceptable;

d) It generates living beings whose sentience has been reduced to an excessive extent.

Whilst points a) and c) in the above should be adequately catered for under current welfare regulations, decisions about unacceptable violation of integrity or reduction in sentience are not.

FAWC's Cloning Report commented on potential problems concerning violation of integrity or unnaturalness which, in the absence of suitable controls, might well result in a significant insult to the animals involved. We stated that we shared concerns expressed to us in the consultation exercise that "an attitude may be developing which condones the moulding of animals to humankind's uses, irrespective of their own nature and welfare". In the case of cloning, this was a perception of a cloned animal as a manufactured being, which to some in society is offensive. We also stated that, "it is not clear that a radical distinction between human and non-human is now defensible, either biologically or ethically, nor that any such disjunction is sufficient to warrant the treatment of other living creatures merely as means. We owe respect to other animals, and especially to those which we choose to domesticate."

Both conventional and novel breeding techniques have the capacity to produce animals whose integrity has been altered to an unacceptable degree. The problems of addressing this complex question have already been considered in the section 'Gaps in current regulations' (paragraphs 41-59), where we stated that we support the recommendations of the Banner Committee and the AEBC that there is a need for consideration to be given to these questions. The following examples further illustrate the type of problems that need to be considered.

An example of a possible candidate for such ethical consideration is the featherless broiler chicken, produced in Israel by conventional breeding methods. Such an animal might not be excluded from commercial production on welfare criteria since it is feasible that the environment for which it was selected may actually favour baldness. However, it might be argued that such a significant change to genotype or phenotype should be prohibited from entering commercial production on the grounds that it constitutes an intrinsically objectionable change to the nature or 'integrity' of the animal.

Another example where a broader set of ethical considerations, rather than a purely welfare based approach, might be required is for the commercial acceptability of a strain of laying hens that are "genetically blind". Researchers in Canada concluded that when compared with sighted hens, the blind birds laid more eggs, consumed less food, were less affected by flock size and stocking density, and had better feather cover. The researchers suggested that on the basis of their evaluation of welfare, the blind birds may have reduced stress levels and that it was worthwhile to explore further the potential of this mutation in egg-laying strains kept in cage systems.

A final issue, briefly addressed in the section 'Gaps in current regulations' (paragraphs 41-59), is that of selecting animals for behavioural traits. A reduction in sensitivity to the environment is a general effect of domestication in many species, but FAWC is aware that selection for temperament is becoming increasingly important in breeding programmes. This is particularly the case for species such as pigs and laying hens, where a move away from close confinement systems, driven by either legislation or market forces, has revealed the importance of behavioural traits such as reduced levels of aggression. Whilst breeding for temperament has been carried out for hundreds of years, the protection of behavioural flexibility and sentience in animal breeding is becoming an issue where regulation may be necessary.

The above examples demonstrate the wide range of issues that demand proper ethical evaluation on the basis that they constitute major changes to the integrity or sentience of animals. For simplicity, we have chosen not to address the possibility that these examples pose a more obvious risk to welfare, for example, that 'blind' chickens are more efficient because they are less active.

Proposal for a Standing Committee to Consider Animal Breeding in Agriculture

Against the background of the problems identified by FAWC throughout this report, we recommend a vehicle through which the majority of these can be addressed:

Recommendations

FAWC recommends that a Standing Committee be established for the evaluation of new and existing breeding technologies as well as for the consideration of welfare and ethical problems arising as a result of livestock breeding programmes.

FAWC recommends that the Standing Committee provide advice to Government on the effectiveness of existing legislation, and the possible gaps that exist, relating to farm animal breeding procedures, in order to promote animal welfare.

FAWC recommends that the Standing Committee give due consideration to ethical questions associated with animal breeding even where measurable detrimental effects on animal welfare may not be immediately evident.

FAWC recommends that any breeding technology, whether developed within the UK or overseas, be thoroughly evaluated by the

Standing Committee prior to, and during, its incorporation into commercial agricultural practice in the UK.

Proposed Model for the Standing Committee

FAWC has considered carefully whether the proposed Committee would operate most efficiently and appropriately within or outside the umbrella of FAWC. Given the obvious overlap in subject and remit, FAWC's international recognition for its advice on farm animal welfare, and the good working relationships that the Council has with many sectors of the UK livestock industry and welfare interest groups, we are of the view that it would be appropriate and advantageous for the proposed Committee to be closely linked to FAWC in the form of a Standing Committee of FAWC. However, the membership would be largely, if not entirely, appointed from expertise outside the membership of FAWC.

FAWC's typical working method is to focus on a specific topic and produce a report and recommendations over a period of two to three years for the action of either Government or the food and farming industry. Most of the issues raised in this current report would require a very different approach, as progress would only be made as a result of long term programmes of work, often in close co-operation with breed companies. The resource issue that this additional working practice would present, in addition to the specialist expertise that the Committee would require, is a key reason why the existing members of FAWC would be unable to fulfil the role and remit of the proposed Committee, and therefore why a Standing Committee with a separate membership is essential.

To strengthen the link between FAWC and its Standing Committee, the Chairperson of the Committee should be appointed to FAWC and should, as with all members of the Standing Committee, be appointed by Ministers. Advice to Ministers from the Standing Committee would normally be provided via FAWC. Thus, all reports from the Standing Committee would normally be endorsed by FAWC and presented to Ministers as FAWC Reports.

The Standing Committee, established and appointed by Ministers, could be known as the FAWC Animal Breeding Committee (FAWC ABC) and should be composed of members representing a broad spread of expertise and interests, together with lay representation and, importantly, an independent chairperson. The composition of the Committee should be wide enough to cover all major areas of interest and also have sufficient in-depth expertise to evaluate the evidence brought before it. For instance, given the Committee's remit, it would be essential to include at least one specialist in molecular genetics in the Committee membership. Additional expertise may be introduced through secondment of experts as members to assist with specific issues. In recognition of the long term issues that the Committee will address, it would be advantageous for the term of appointment of Committee members to be longer than that of FAWC members. This could be, for example, five years.

FAWC proposes that the Standing Committee would work from information from a variety of sources such as Ministers, Government departments, the State Veterinary Service, the Home Office Inspectorate, the VLA, RCVS as well as from FAWC itself. It should have the freedom to determine its priorities for investigation and have the facilities to make in depth studies of new or complex issues.

Given that a major part of the Committee's work would be associated with the surveillance of specific welfare problems, it would be essential to establish appropriate mechanisms to provide this information, such as the collection of good quality data that allows analysis of genetic and environmental effects on important health and welfare traits. For some species this data may in effect be already available, for example, data on lameness and somatic cell count are collected in dairy cattle and have been analysed to examine the respective contribution of genotype and environment. However, for other species it is likely that a greater range of traits requires to be measured and steps would need to be taken to incorporate these in existing measurement schemes.

Where available, existing data should be utilised. However, where new information is required, this should be obtained through the promotion of industry partnerships and utilisation of research procurement mechanisms already in existence within Government. The obvious overlaps between the requirement of the Committee for farm-level data and the development of the Government's Health & Welfare and Veterinary Surveillance strategies should also be acknowledged.

Recommendations

FAWC recommends that targeted surveillance is made of farms where new breed types or new breeding technologies are first introduced into commercial practice, and that the welfare impact of all such developments is reviewed throughout a period of normally not less than 5 years after introduction into commercial agriculture.

In order to determine the consequences of current breeding strategies or any new breeding technology and to provide essential feedback on welfare performance for breed companies, FAWC recommends that a robust surveillance system be established. This should accurately monitor the incidence of specified on-farm welfare problems and be capable of providing information on welfare problems associated with breeding strategies or technologies and to determine the respective genetic and environmental contributions. This surveillance system should include extensive data currently collected, for example, by breed societies and breed companies, and should be developed in association with, and as part of, the Government's Animal Health and Welfare and Veterinary Surveillance Strategies.

As far as possible, the Standing Committee will be required to work in an open and transparent way, recognising however that certain issues may be subject to strict commercial confidentiality. The issue of confidentiality could be overcome by negotiation of binding confidentiality agreements as already exist between breeding companies and academics involved in analysing their data.

The Standing Committee would be required to operate in a manner which would not place a disproportionate bureaucratic burden on agriculture, although it is important to recognise that some advice will be far from straightforward and would involve in depth discussion with experts in particular fields. In many cases, the Standing Committee would advise the continued field surveillance of a particular genotype or technique before making a final recommendation.

Given the considerable public concern over breeding technologies particularly within areas such as GM and cloning, it would be essential for the proposed Committee to be perceived by the public as a responsible voice on such issues, in much the same way as the Human Fertilisation and Embryology Authority (HFEA) or, in New Zealand, the Bioethics Council. For this reason the Committee should be demonstrably independent and should also have a role whereby it engages in a two-way dialogue with the public.

The Standing Committee will require the support of an effective, adequately resourced secretariat and must have access to resources to enable research and surveillance to be commissioned for particular issues under consideration.

Whilst FAWC has considered breeding related welfare problems in farm animals, consideration could be given to expanding the role of the proposed Standing Committee to cover similar problems found within companion animals. In this way, the Committee could engage with CAWC in addition to FAWC.

Concluding Remarks

FAWC has considered carefully the options available for addressing the broad range of ethical and welfare issues that relate to breeding and the application of breeding technologies in farm animals, as raised in this report.

We believe that any failure to address the issues highlighted presents a significant risk to Government, the livestock industry and, most importantly, to animal welfare. For example, there is considerable public disquiet about genetic modification, cloning and some novel breeding technologies. At the present time it is difficult to predict the extent to which developments in these fields will become incorporated into livestock breeding programmes. However, it seems reasonable to assume that public opinion will be an important factor influencing developments in these areas. A crucial role of the proposed Standing Committee would be to be seen by the public as a trusted and reliable body to provide balanced advice to Government and at the same time to listen to public concerns about such matters. In addition to helping avert potential risks, the proposed model for a Standing Committee would provide a number of other benefits. For example, analysis of data from an effective welfare surveillance system would provide information on both genetic and environmental influences on health and welfare, thus allowing both aspects to be addressed in a coherent way. Such a welfare surveillance system would also allow welfare problems to be addressed when they first become apparent and not, as is often the case, many years after they develop.

In the same way, research effort in many areas of farm animal welfare could be much more effectively and carefully targeted if accurate data on the prevalence of welfare problems were available. The proposed welfare surveillance system would, in itself, be a resource of enormous value in that it would allow trends in a wide range of welfare problems to be monitored, thereby assisting Government and other interested parties (e.g. research groups) to focus attention on the most pressing problems. At the same time it would allow industry to demonstrate where recognised welfare problems were being addressed, both through selective breeding as well as through management. 136. Animal breeding and the use of breeding technologies is a dynamic and growing field that has the potential to influence animal welfare in a positive, as well as negative, way. The proactive approach recommended by FAWC to address the issues raised in the report would ensure that neither progress nor welfare are compromised.

Comments and Recommendations from Previous FAWC Reports

Report on the Welfare of Broiler Chickens, April 1992

It is apparent that there are many possible remedies for leg problems but none which will alone solve the entire problem. Therefore, we look to the industry to take steps to tackle this matter. Principal methods of reducing leg problems are outlined here for the guidance of those concerned:

Genetically – by the increased selection of breeding stock for strong and well-formed legs. The council acknowledges that the industry is already working towards a reduction in the predisposition of broilers to leg problems. However, this change is too slow and long-term.

Having considered all aspects of this problem, including it's long-term nature and the attempts already being made by the industry to improve the situation, *the council concludes* that the current level of leg problems in broilers is unacceptable. We recommend that steps should be taken to ensure that there is a significant reduction in the numbers and severity of leg problems. It will be the responsibility of the industry to achieve this objective and the council intends to look at this aspect of broiler production again in 5 years' time, when significant improvements should be apparent. If no reduction in leg problems is found, we may recommend the introduction of legislation to ensure the required improvements. To achieve these improvements further R&D may be necessary. Genetic selection has the potential for positive as well as negative effects on welfare. However, the selection of stock for liveweight gain and food conversion efficiency in preference to, and to the detriment of, factors necessary for the welfare of the birds should be discouraged. The selection of stock for the reduction of leg problems is strongly encouraged.

Report on the Welfare of Sheep, April 1994

Traditional breed structures have been under change in recent years in response to the demand for 'improved' leaner carcass conformation. We are concerned that such changes are at the expense of the welfare of the sheep involved, particularly in extensive environments.

We recommend that if any change in breed or type is contemplated in difficult extensive conditions, replacement must only be with a breed or type which is sufficiently hardy. Account must also be taken of the effect of introducing breeds which are unfamiliar with area.

We recommend that within breed improvement selection programmes, monitoring is carried out for problems associated with selection for greater muscularity.

The body condition of the ewe and nutritional management around mating time are well proven to have a marked effect on the ovulation rate and thus litter size. In addition, there are other methods of manipulating time of lambing and litter size. This is an obvious opportunity for the use of the scanning technique. The use of artificial melatonin treatment of the ewe may be used to induce reproductive activity about 6 weeks earlier than normal.

We recommend that if any procedure to increase litter size is instituted, account must be taken of the extra requirements in welfare, feed, labour and other inputs before and at lambing time.

We also recommend that appropriate housing or shelter should be available if lambing is to take place outside the normally recognised breeding period in adverse weather conditions.

Many sheep farmers use cervical artificial insemination as an aid to genetic improvement. There has, however, been a recent development whereby semen is deposited directly into the uterus beyond the cervix. We believe that the difficulties associated with the technique of transcervical A.I. mean that it should not be carried out by anyone other than a veterinary surgeon. Embryo transfer is a technique only for use by approved teams led by a veterinary surgeon.

We *recommend that only a qualified veterinary surgeon trained in the technique should* perform transcervical A.I. in sheep.

Report on the Welfare of Turkeys, January 1995

Large breeding stags which are used as grandparent stock to influence the genetic pool have the greatest potential for leg disorders. There are very extreme types with a high predisposition to lameness. Very few breeding stags are taken to excessively heavy weights but those we saw had unnatural gaits and some had difficulty in walking.

Breeding policies adopted across the world make it inevitable that a small number of grandparent stags will be reared to extreme weights and size. These birds should not be allowed to suffer. We know of evidence that the older, heavy breeding stags may be affected by severe degenerative joint disease and experience pain during movement as a result. Further research is required to establish whether or not these lesions occur in younger breeding stock, particularly males; and to determine the potential for suffering and whether or not the lesions are related to excessive weight or abnormal conformation. In the meantime, management practices must seek to minimise the occurrence of leg pain in heavy stags by careful selection, vigorous culling and reduction of the energy intake levels. Culling should occur immediately a problem is manifest and breeding companies must not continue to collect semen once a stag shows signs of suffering.

We believe that some grandparent stags have been taken beyond the limit of acceptability and that it would not be appropriate to continue to select for increased weight without first improving leg strength.

We recommend research to establish at what point leg problems, particularly in heavy breeding stags, result in pain or other aspects of poor welfare and into the predisposing factors and pathological lesions which may be associated with leg disorders.

"Shaky leg syndrome", which appears to be the commonest leg disorder of turkeys, is not characterised by any distinctive pathology. The condition requires further research.

Farmers and breeding companies must rigorously cull birds which are suffering as a result of leg disorders.

Heavy grandparent birds should not be kept to advanced ages.

Report on the Welfare of Pigs Kept Outdoors, May 1996

The breed of pig selected by the farmer must be suited to outdoor production and it is not acceptable simply to turn out sows which have spent their lives indoors should the farmer decide to convert to outdoor production. The pig should be of a hardy type, able to cope with most climatic conditions in this country, and with good mothering instincts. The breeding companies have reacted to market demands and have developed genotypes which are both hardy and which produce a carcass quality and production efficiency which compares well with pig meat produced indoors.

It is the practice in the UK to select relatively docile sows for outdoor production on the basis that they are easier to manage and have good mothering instincts. As a result, injury arising from aggression between animals is relatively rare. The large space available for avoidance and escape, together with proper management attention, also help to reduce aggression. The protective instincts of sows with strong mothering ability should be recognised as being distinct from general aggression towards other stock.

Breeding companies, and those responsible for the selection of breeding stock to be kept on outdoor enterprises, must ensure that only those strains of pig with the genetic potential to thrive in the conditions provided are used.

When choosing pigs attention should be paid to the need for good temperament and mothering ability. New entrants to the industry should seek independent advice to ensure the correct stock is selected.

Report on the Welfare of Laying Hens, July 1997

We are greatly concerned to hear of an increase over recent years in the incidence of beak trimming of hens in battery cages. This increase may be associated with genetic change in the laying flock. It may also be due to the fact that many hatcheries routinely beak trim since birds may be destined for any system. It seems possible that rather than exploiting genetic variation in feather pecking and cannibalism to reduce welfare problems, the breeding companies may have inadvertently made the situation worse. Evidence from the USA that genetic selection can significantly and substantially reduce feather pecking and cannibalism is encouraging. There appears to be great potential for genetic selection to overcome, either partly or wholly, the problem of feather pecking and cannibalism and hence the need for beak trimming. We would like to see evidence of a concerted effort by the egg industry and retailers to ensure that the laying hen breeding companies reduce the genetic tendency of the hen to injurious feather pecking and cannibalism.

In addition, all breeding companies should be strongly encouraged by the laying hen industry and Government to pursue genetic selection for birds which display less injurious behaviour.

Achievement of good welfare should be of paramount importance in breeding programmes. Breeding companies should devote their efforts primarily to selection for health traits so as to reduce current levels of lameness, mastitis and infertility; selection for higher milk yield should follow only once these health issues have been addressed.

Breeding programmes worldwide should have as a major objective the need for good welfare. The criteria used should not lead to the production of animals that require above average levels of management to prevent welfare problems.

Subject to the recommendations at paragraphs 40 and 41 of the main text, breeding companies should pay special attention to the selection of cattle with good feet and legs so as to reduce the risk of lameness. This may require the collection and processing of more on farm data to aid better choice of breeding animals.

Sire selection should take account of good linear assessment scores for foot depth and angle and straightness of legs. Replacements should not be bred from cows or sires with a history of severe clinical lameness or badly deformed feet and legs.

Bulls should be culled if their offspring have a poor record of foot health and conformation, even though they may be genetically superior in other traits.

In recent years, the desire to increase rapidly the genetic production index has led to a growing acceptance in the industry of embryo transfer. The procedure is currently undertaken on a relatively small, albeit increasing, scale.

The procedure in cattle is mainly carried out by superovulation and non-surgical recovery and involves the transfer of embryos which may have been fertilised *in vivo* or *in vitro*. Ovaries may also be collected from dead animals in the abattoir and the ova grown-on and fertilised in the laboratory before transfer. These embryos may be transferred directly or frozen for storage and future use. The procedure for transferring single embryos to carefully selected recipients does not normally cause welfare problems. The continued use of superovulatory drugs can result in subsequent fertility problems.

Embryo transfer has now been simplified with the advent of new knowledge which allows frozen embryos to be thawed very quickly and transferred immediately to recipient females. As embryo transfer is carried out at seven days after the onset of oestrus, the technique is more difficult than AI and requires considerable training and experience. Caution must be exercised if this practice is to become widely available in the commercial field, as embryos fertilised *in vitro* have been implicated in the production of oversize calves. The technique must be carried out using epidural anesthesia. If a non-veterinarian is administering the epidural, that person must be trained and certified as competent and be a member of an approved bovine embryo team headed by a veterinary surgeon.

These procedures are regarded by the industry as part of the normal array of reproductive techniques and welfare concerns are, by and large, addressed by The Bovine Embryo (Collection, Production and Transfer) Regulations 1995 and the Royal College of Veterinary Surgeons' Code of Practice for Embryo Collection and Transfer in Cattle.

Invasive techniques of this nature should not be undertaken lightly. In particular, the transfer of embryos which are likely to produce calves of a size or shape which will cause problems at calving, or increased likelihood of caesarean section, is unacceptable. We have recommended previously that the Belgian Blue should not be used on maiden heifers, either as semen or by embryo transfer.

The effects of repeated administration of superovulatory drugs and repeated epidural anaesthesia should be kept under review. The industry should maintain records of problems caused and report findings annually to the Agriculture Departments.

The requirements of the Bovine Embryo (Collection, Production and Transfer) Regulations 1995 must be carefully adhered to at all times.

The recipient mother should be able to carry the chosen embryo to term and to calve normally, without recourse to caesarean section. Caesarean section must not become a routine part of the procedure.

A record of caesarean sections resulting from embryo transfer should be maintained and submitted annually to the Agriculture Departments who should monitor the situation.

Embryo transfer which does not fulfil these criteria should be regarded as causing unnecessary pain and unnecessary distress.

We have considerable concerns regarding oversize calves resulting from in vitro fertilised embryos that can cause difficulties at calving. It is essential that research is pursued as a matter of urgency. We intend to review the evidence within five years of this report.

Repeated epidural injections to allow for collection of oocytes by follicular aspiration as frequently as twice a week can cause welfare problems, such as chronic pain in the tail head and fused vertebrae, and requires urgent study. Pending these results, the frequency of ovum pick-up should be limited. The industry should be required to maintain records of problems caused and to report its findings annually to the Agriculture Departments.

When commercially available, the sexing of semen should be used to reduce the number of unwanted male dairy calves, provided that the technique has not been shown to produce adverse effects.

Report on the Implications of Cloning for the Welfare of Farmed Livestock, December 1998

We have considered carefully the welfare implications of the procedures involved, and have concluded that it is most important that *in nuclear transfer, the recipient dam must be of* an appropriate size in relation to the size of offspring to be produced. Regulations should be introduced requiring the suitability of the dam to be certified by a veterinary surgeon in a manner similar to that required in embryo transfer.

From the summary of recommendations:

— We have identified three key areas where we believe that greater knowledge and understanding is required. We consider that:

   * *a greater understanding is required of the underlying causes of oversized offspring,* especially of the effects of in vitro culture on foetal size;

   * *research is needed into the causes of embryonic and foetal deaths and of perinatal* losses and birth abnormalities associated with nuclear transfer, and the scale of these problems; and

   * *research into the long-term effects of nuclear transfer of aged DNA is required before* the technique should be considered suitable in agricultural practice.

— Bearing in mind this requirement for additional information, we make an overriding recommendation that, until the problems of oversized offspring, embryonic and foetal losses and birth abnormalities, and the possibility of problems associated with aged DNA, have been satisfactorily resolved, there should be a moratorium on the use of cloning by nuclear transfer in commercial agricultural practice.

* We were concerned to learn that, in the current practice of nuclear transfer in sheep, oocytes are cultured in vivo which involves the accrued stress of a surgical procedure with recovery followed by killing of some animals. Whilst we accept that the number of animals involved is small, we nevertheless believe that *surgical intervention of animals used for in* vivo culture with subsequent killing should be avoided. Further research is required to minimise stress in, and avoid wastage of, such animals. Research should also be undertaken to develop in vitro culture methods.

* Furthermore, in addition to the need for greater understanding and knowledge, we have identified the need for a new regulatory structure and therefore recommend that, *before* cloning is permitted in commercial agricultural practice, legislative controls must be introduced. These must include:

a) regulations to give protection to cloned farmed livestock similar to that enjoyed by research animals, at least until the effects of the cloning and any associated genetic manipulation have been scientifically evaluated in the environment of commercial agricultural practice.

b) regulations to ensure the procedure is carried out under direct veterinary supervision by adequately trained personnel.

A further aspect of good welfare lies in controlling the competence of those who carry out procedures. We therefore recommend that *the Royal College of Veterinary Surgeons should* be consulted to explore the feasibility of any of the procedures involved in cloning by nuclear transfer which are "acts of veterinary surgery" being suitable for delegation to a trained lay operator who has attended an appropriate course and has been assessed for competency. We also have recognised the importance of good stockmanship and recommend that stockpeople responsible for the care of animals which arise from cloning must be sufficiently trained and competent to attend to any particular requirements of such animals. We consider that loss of genetic diversity may have a deleterious impact on welfare. Consequently, we recommend that *an effective system of control of cloning by nuclear* transfer, or similar means, must be implemented to ensure that genetic diversity is maintained and other adverse effects prevented. Such control may need to be statutory and should ensure that proper breeding records of any animal produced by such cloning are maintained for several generations.

We consider that implementation of our recommendations will extend over a period of several years during which emerging information will need to be evaluated. We therefore recommend that a *National Standing Committee should be established to oversee the* development of cloning technology. The Committee should review outputs of research aimed at tackling the welfare problems identified in this report (and any other problems which may emerge); it should determine the time when it may be appropriate to introduce cloning into commercial agricultural practice; and it should ensure that the controls by then in place are both adequate and effectively implemented.

We also recognise that the advances in this and similar technologies are taking place around the world and there is a need for international guidelines. We therefore recommend that *liaison at an international level should be established to ensure that similar controls are* in place wherever this technology is being applied. Finally we believe there is a need, as a matter of urgency, for a two-way exchange of information on this and related technologies. We therefore recommend that *a means of* exchange of information on animal cloning and possibly related technologies, should be put in place as a matter of some urgency. The intent should be both to improve public awareness of the facts and issues, as well as to improve politicians' and scientists' understanding of the fundamental public concerns which undoubtedly exist. Participants should include representatives from the relevant industries and academic groups, as well as politicians and the range of public interests.

Interim Report on the Animal Welfare Implications of Farm Assurance Schemes, August 2001

We consider that other wider aspects of scheme standards require further attention and research. *We recommend that consideration be given, in particular, to the incorporation* of scheme standards, which relate to the breeding and rearing of animals for specific production systems, and also the treatment of cull animals at the end of their productive *lives.* position on a chromosome.

3

Genetic Engineering and Traditional Breeding Methods: A Technical Perspective

The Soil Association's rejection of the use of genetic engineering (GE) in agriculture as simply having "no place in organic food and farming" (Living Earth, Jan. '97), is justifiable purely as a matter of principle. GE represents an extension of intensive, industrial agriculture and therefore reinforces environmentally damaging, non-sustainable husbandry. Evidence already exists which demonstrates that the claims that GE crops will result in less dependence on agrochemicals are, in the medium to long term, unfounded.

The greatest claim of those who endorse the use of GE in agriculture, is that it is a safe, more precise and natural extension of traditional cross breeding methods for generating novel varieties of crops and farm animals. It is said that this new technology simply gives nature a helping hand with something that would happen anyway. The aim of this article is to assess GE in agriculture from a technical and basic genetics viewpoint focusing in particular on plants and animals. We will see that technically speaking, the use of GE in agriculture is a crude and imprecise technology which bears no resemblance to traditional breeding methods for producing new varieties of crops and farm animals. Given this imprecision, the outcomes of using GE in food production both in terms of potential ill health and negative environmental impact, are far from certain. There would therefore also appear to be good scientific grounds for questioning the validity of using GE in agriculture especially when there are safe alternatives available.

The Fundamentals-Genes and Genetics

Genes are discrete units of DNA. They are the blueprints which carry the information for the tens of thousands of proteins which act as the building blocks of all the structures and functions (biochemistry) that constitute the body of any organism from bacteria to humans. DNA can be likened to a long string of pearls where each pearl, representing a gene, occupies it's own special place in the "necklace" which is vital for it's correct function. Genetics, the study of genes, has two basic components. Firstly, there is the information content of each gene; that is, what gene carries the blueprint for which protein. Secondly, genetics has taught us that the activity or expression of each gene is extremely tightly controlled or regulated. Put simply, each gene has it's own set of sophisticated on-off switches to drive it's expression ensuring that the correct protein and therefore appropriate structure and function, is present in the right place, time and quantity in the body.

Just as all forms of life are interdependent upon each other for survival and growth, no gene works in isolation from all other genes. *The latest discoveries tell us that genes are arranged along the DNA in groups or "families". The function of a given gene in a group is dependent on all the other genes that are present within the same family. Furthermore, the genetic activity in one family of genes can effect the function of genes in other groups of genes. It is also clear that genes and the proteins that they give rise to, have co-evolved together to form an extremely intricate, interconnected network of finely balanced functions the complexities of which we are only just beginning to understand and appreciate.* Such tight control of gene activity means you will never find liver functions in your brain or leaf specific processes in the fruit and vice versa! In addition, Nature has also evolved mechanisms whereby cross breeding can only take place between very closely related species. With traditional breeding methods, different variations of the same genes in their natural context (within the necklace of pearls) are exchanged. This preserves tight control and complex interrelationships between genetic and protein functions that are vital for integrity of life as a whole.

GE: A Natural Extension of Traditional Breeding Methods?

In order to assess the validity of the claim that GE represents a natural extension of traditional breeding methods, it is important to know how GE ("transgenic") plants and animals are produced.

GE Plants

As an example, let us see how the herbicide resistant, GE soya was generated. The objective here was to introduce into the soya plants a gene from a common soil bacterium which would allow it to survive when sprayed with the herbicide Roundup. Clearly you cannot "cross" a bacterium with a plant.

Therefore, the first step was to grow cells from soya bean plants on plastic dishes in the laboratory. Now, in order to allow the bacterial gene to be able to work once introduced into it's new plant host, it had to be linked to a genetic switch combining parts from a cauliflower virus and petunias. (As we discussed above, the bacterial gene's own switch will only work in the bacteria from which it came). This combination of cauliflower virus, petunia and bacterial DNA was then introduced into the soya bean cells growing on the dishes in the laboratory using a procedure known as "biolistics" which employs a device called a "gene gun". In this technique, tiny spheres of gold or tungsten are coated with the DNA one wishes to introduce into the plant cells. These DNA-coated metal particles are then shot at the plant cells using the gene gun at high speed. As a result some of these metal beads enter inside the plant cells carrying the new DNA with them.

Unfortunately from the point of view of the plant biotechnologist, the efficiency with which the new DNA is taken up by the soya bean cells on the dish is very low. Most of the cells don't take it up at all. So the key is to find those few cells among the many millions on the dish which have taken up the DNA. This is done by using another genetic trick. The introduction of the bacterial gene into the soya bean cells for herbicide resistance, was accompanied by a second gene which confers resistance to an antibiotic (called kanamycin). The soya bean cells were then treated with the antibiotic. The few cells which had taken up the herbicide resistance: antibiotic resistance "marker" gene combination survived and flourished whereas the majority of the cells which had not taken up these genes were simply killed by the antibiotic.

Finally, by changing the conditions under which the soya bean cells are grown, the cells clump together to form what is called a callus which in turn starts to put down roots and sprout green shoots. These little "seedlings" are then potted so as to grow into fully mature plants which will carry in all their cells (including those for reproduction;

i.e. pollen etc.) the new bacterial gene. The plant which then displays the best agronomic performance, in this case resistance to herbicide, is then selected for further development (crossing to form new hybrids etc.).

GE Animals

The generation of transgenic animals is a somewhat simpler, but no less artificial procedure. Fertilised eggs are first removed from the animal of choice. These eggs are then injected with the genes one wishes to engineer into the animal. The DNA injected eggs are then returned to the womb of a surrogate mother where they complete their development and are born in due course.

Therefore, in marked contrast to traditional breeding methods, all transgenic plants and animals start life as individual or groups of cells growing on a plastic dish in a laboratory.

GE: A No Holds Barred Technology

It is evident from the procedure we just described that with GE there are no holds barred. GE allows the isolation, cutting, joining and transfer of single or multiple genes between totally unrelated organisms circumventing natural species barriers. As a result combinations of genes are produced that would never occur naturally. Transgenic crops containing genes from viruses, bacteria, animals as well as from unrelated plants have been generated.

In the case of the herbicide resistant soya beans, the final outcome was the combination of genetic material from four totally unrelated organisms; a cauliflower virus, petunia, bacteria and soya. Furthermore, again as we saw in the case of the GE soya beans, the newly introduced gene units are composed of artificial combinations of genetic material.

Another example which illustrates the extreme combinations of genetic material that can be produced, is the introduction of the "anti-freeze" gene from an arctic fish (the sea flounder) into tomatoes, strawberries and potatoes in the hope of producing resistance to frost.

As with the bacterial gene in the soya beans, the fish anti-freeze gene is joined to the cauliflower virus genetic switch to allow it to turn on and work in it's new host. (The fish genetic switch naturally only works in the fish). All this is in turn coupled to an antibiotic resistance marker gene to allow selection of the newly transformed plants.

GE Disrupts Host Gene Functions and Possesses Inherent Unpredictability

Clearly GE represents a great technological advance. However, as we have already discussed, genes have evolved to exist and work in families. Therefore, the claim that the reductionist approach of GE which moves one or a few genes between unrelated organisms, is a precise technology is highly questionable.

Furthermore, the generation of transgenic plants and animals is currently an imperfect technique. Once injected into the cells of the organism, the introduced gene is randomly incorporated ("spliced") into the DNA of it's new plant or animal host.

In fact, the manner in which GE animals and plants are produced, always selects for the splicing of the foreign gene into regions of the host DNA where other natural genes are trying to work. Given the interdependence of gene function within any grouping of genes, this random splicing of the foreign gene into the host DNA will always result in a disruption in the normal genetic order in the "string of pearls". Therefore, GE of animals and especially plants, always results in a loss, to a lesser or greater degree, of the tight genetic control and balanced functioning which is retained through conventional cross breeding. With GE, host genes can be silenced (inactivated) or inappropriately switched on resulting in either a deficiency in a given protein(s) or the presence of the wrong protein(s) in the wrong place or in the wrong quantity or all these combined.

In addition, it is also assumed that the introduced gene and the protein that it makes, will behave in exactly the same way in it's new host as it does in it's native environment which frequently will not be the case. As discussed above, gene and protein functions have evolved over millions of years to work together in any given organism. The anti-freeze gene/protein in the arctic sea flounder has evolved to work together with the other genes/proteins in this fish. It is purely an assumption that it will work in exactly the same way with no unwanted side effects in it's new hosts where it will now be surrounded by plant proteins!

These effects combine to produce a totally unpredictable disturbance in host genetic function as well as in that of the introduced gene. The resulting disturbance in biochemical function can unexpectedly produce novel toxins, allergens and reduced nutritional value.

Conclusion: GE and Traditional Breeding Methods Are Worlds Apart

The proponents of the use of GE in agriculture argue that mankind has been selecting and manipulating plant and animal food stocks for millennia and that this new technology is simply the next stage in this process. However, we have seen:

- Technically speaking, GE and traditional breeding methods bear no resemblance to each other.

- GE plants and animals start out life in a laboratory culture dish.

- GE employs totally artificial units of genetic material which are introduced into plant and animal cells using chemical, mechanical or bacterial methods.

- GE always results in disruptions to the natural order of genes within the host DNA.

- GE also brings about combinations of genes that would never occur naturally.

Clearly these procedures are worlds apart when compared to cross fertilisation between closely related species. The totally artificial nature of GE does not automatically make it dangerous. It is the imprecision in the manner by which genes are combined and the unpredictability in how the introduced gene will interact within it's new environment which results in uncertainty. The balanced gene functions that have evolved together and which are preserved with traditional methods, are lost with GE. Therefore, from the standpoint of the fundamental principles of genetics and the limitations in the technology, GE is neither more precise nor a natural extension of traditional cross breeding methods. If anything the opposite would appear to be true. Therefore GE foods possess new and unique safety considerations both in terms of health and to the environment.

The availability of safe, sustainable, natural methods of breeding and husbandry utilising the many thousands of different varieties of a any given food crop, makes the risks associated with GE foods simply not worth taking. These risks are even less acceptable when one takes into account the fact that once released into the environment, genetic mistakes/pollution cannot be recalled, cleaned up or allowed to decay like agrochemicals or a BSE epidemic, but will be passed on to all future generations indefinitely.

Breeding and Supply

Many animal models of human disease are chosen because they may be readily bred in captivity, enabling many generations to be evaluated in a relatively short period of time. Genetically altered (GA) animals are also important, often as models of specific human disease. The links above provide information on breeding of the most commonly used vertebrate species, including GA mice. Although invertebrates such as the nematode worm *Caenorhabditis elegans* and the fruit fly *Drosophila melanogaster* are important to biomedical research, a discussion of breeding techniques of invertebrates is beyond the scope of this Information Portal.

Many laboratory animals are obtained from commercial breeders which are able to supply a wide range of well-defined, high quality animal models representing most of the commonly used laboratory species. In-house breeding is largely confined to the stocks and strains of animal models that are not commercially available, the production of most transgenic animals that now comprise a significant and increasing proportion of those used in biomedical research, and/or to the maintenance of colonies of animals associated with studies of reproduction and genetics.

Breeding methods which are suitable for the production of animals from in-house colonies are outlined in Festing and Peters (1999). Details of genetic considerations in the maintenance of inbred, outbred and mutant rodent stocks are discussed in Festing (1999), and methods of refinement and reduction in the production of GA mice are set out in Robinson et al. (2003). The report of the RSPCA Resource Sharing Working Group gives advice on archiving and sharing of GA mice lines. Although techniques will differ from species to species, some basic principles apply to the breeding of all laboratory animals:

Reproduction and reproductive performance may be influenced by a variety of extrinsic factors and consequently the physical environment is important for maintaining the health and welfare of breeding animals, and for enhancing their breeding performance. Optimal ranges for environmental conditions, as well as minimum standards of husbandry and care, are set out in the Code of Practice for the Housing and Care of Animals in Designated Breeding and Supplying Establishments.

As breeding animals are typically maintained for longer periods than animals used on scientific procedures, particular attention is needed to ensure that the environment provides for the animals behavioural as well as physiological needs. The process of breeding laboratory animals can involve the thwarting of natural behaviours, e.g. laboratory animals may be weaned and separated from their dam at a time which rarely coincides with the time they would have dispersed naturally, and this can be stressful for both the juvenile animals and the parent. It is also important to bear in mind that the needs of infants and juveniles may be different from those of adults.

In situations where breeding stock and/or post-wean stock are housed in groups, individuals that may be disadvantaged in the social hierarchy, e.g. subordinates and females that have recently given birth, may be vulnerable to social stresses. Extra care should be taken to prevent and monitor aggression and to separate individuals if necessary. Reproduction is an energy intensive activity, so animals in poor condition should not be used for breeding.

Breeding animals should be provided an appropriate diet which should be formulated to satisfy their nutrient and energy requirements, particularly for the demands of pregnancy and lactation, and be palatable and free from chemical and biological contamination.

Health status has a significant influence on animal welfare and the use of healthy animals is regarded as a prerequisite both for good welfare and for good science. Although animals may be bred for use as specific disease models, intercurrent disease within the population may call into question the validity of information obtained from scientific procedures and make interpretation of results difficult or even impossible. It is important, therefore, to have a system for monitoring animal health and to have plans to maintain and deal with potential problems, such as disease outbreaks. FELASA has published some recommendations, relating predominantly to the monitoring of microbiological status, for laboratory animal health screening programmes.

Research animal models will invariably be chosen on the basis of particular genetic and/or phenotypic characteristics. When breeding animals for research, therefore, appropriate resources should be directed towards monitoring these essential traits and in these and other ways, maintaining the genetic and phenotypic integrity of the colonies.

Regular handling will accustom young animals to human contact, making future handling and restraint for procedures less stressful for the animal and easier for the handler. For the larger species, attention should also be paid to appropriate habituation and training so as to prepare the animals for their life on study.

Matching Supply and Demand

It is essential to try and ensure that the supply of animals does not exceed the research requirement in order to avoid the generation of an unnecessary surplus and the potential wastage of animals lives. This requires good communication and coordination between breeders, suppliers and users (which is more feasible with in-house breeding), and also within and between multidisciplinary research teams. It also requires good production planning.

A LASA task force has produced a report on the production and disposition of laboratory rodents surplus to the requirements for scientific procedures, which makes recommendations for minimising avoidable excess. The MRC has issued codes of practice for the supply of rodents and supply and use of aquatic species in research.

In most cases, especially with small rodents, supply and demand cannot be balanced exactly. For example, there is often a marked difference in the requirement for either males or females. This generates an inevitable biological surplus to which must be added a managed surplus arising out of variable management practices and user requirements. Where surpluses do occur it is important to review the causes and take appropriate action to try and prevent the situation recurring.

Rehoming may be an option for species that are kept as companion animals if any surplus cannot be redirected for use in essential procedures. Clearly individual animals should only be rehomed where it is in their best interests, where their welfare can be assured, where adequate resources are available for their long-term care, and with the proviso that they are not suffering or likely to suffer any adverse effects from their use in research.

Whenever animals are euthanased every effort should be made to bank and use their tissues and blood if this will avoid the unnecessary killing of another animal.

Information on the production, supply and usage of animals should be provided to, and considered within, each establishments ethical review process to ensure the fullest use is made of individual animals and that numbers bred are reduced to a practicable minimum.

Hybridization and Genetic Engineering

Animals of domestic origin as well as feral animals sometimes produce fertile hybrids with native, wild animals which leads to genetic pollution in the naturally evolved wild gene pools, many times threatening rare species with extinction. Cases include the mallard duck, wild boar, the rock dove or pigeon, the Red Junglefowl (Gallus gallus) (ancestor of all chickens), and carp. Another example is the dingo, itself an early feral dog, which hybridized with dogs of European origin. Genetic pollution is a serious issue: Living organisms can also be defined as pollutants, when a non-indigenous species (plant or animal) enters a habitat and modifies the existing equilibrium among the organisms of the affected ecosystem (sea, lake, river). Non-indigenous, including transgenic species (GMOs), may bring about a particular version of pollution in the vegetal kingdom: So-called genetic pollution. This term refers to the uncontrolled diffusion of genes (or transgenes) into genomes of plants of the same type or even unrelated species where such genes are not present in nature. For example, a grass modified to resist herbicides could pollinate conventional grass many miles away, creating weeds immune to the most widely used weed-killer, with obvious consequences for crops. Genetic pollution is at the basis of the debate on the use of GMOs in agriculture.

A Genetically Modified Organism (GMO) is an organism whose genetic material has been altered using the genetic engineering techniques generally known as recombinant DNA technology. Genetic Engineering today has become another serious and alarming cause of genetic pollution because artificially created and genetically engineered animals in laboratories, which could never have evolved in nature even with conventional hybridization, can live and breed on their own and, what is even more alarming, interbreed with naturally evolved wild varieties.

Common Sub-divisions of Animal Husbandry

Historically, animal husbandry is broken up into animal subdivisions, agriculturists tending to specialize on one or two types of animals. Below is a list and brief description of some of the major subdivisions:

- Swineherd: A person who cares for hogs and pigs (older English term: Swine).
- Shepherd: Usually a person who cares for sheep. However, in previous years, it was common to have herds which were made

up of both sheep and goats, in which case the term shepherd was still employed.

- Goatherd: Someone who cares primarily for goats.
- *Cowherd/Cowboys: Refers to those who raise, milk, and slaughter cattle. In more modern times, the cowboys or vaqueros of North and South America ride horses and participate in cattle drives to watch over cows and bulls raised primarily for food.*
- Herder: A generic term for anyone who tends to a herd of animals not listed above, such as Camels, yaks, and in Latin America, llamas and alpacas.
- Equestrian: A person who breeds and takes care of horses.
- Breeders: Generic term for anyone who specializes in the breeding of agricultural animals. More often, breeders are trained in specialized techniques, such as artificial insemination and embryo transfer, and are therefore independent hires of the usual farm staff.

An Overview of Animal Breeding

The face of animal breeding has changed significantly over the past decades. Animal breeding used to be in the hands of a few distinguished 'breeders', individuals who seems to have specific arts and skills to 'breed good livestock'. Nowadays, animal breeding is much dominated by science and technology. In some livestock species, animal breeding is in the hands of large companies, and the role of individual breeders seems to have decreased. There are several reasons for this change. Firstly, the breeding industry has taken up scientific principles. Looking was replaced by measuring, and an intuition was partly replaced by calculations and scientific prediction. Other major developments were caused by the introduction of biotechnology. These are roughly the reproductive technologies, and the molecular genetic technology. Not all of this is new. Artificial insemination was introduced in the fifties in cattle.

No doubt that the technology had a major impact on rates on genetic improvement in dairy cattle, and just as important, on the structure of animal breeding programs. Nowadays, technologies like ovum pick up, in vitro fertilization, embryo transfer, cloning of individuals, cloning of genes, and selection with the use of DNA markers are all on the ground. Some of the technologies are already applied, others are further developed, or waiting for application. Finally, the rapid development of computer and information technology has greatly influenced data collection and genetic evaluation procedures in livestock populations, now allowing comparison of breeding values across herds, breeds or countries. The introduction of breeding methods typically needs to find the right balance between what is possible from a technological point of view and what is accepted by the decision makers and users within the socio-economic context of a production system. Ultimately it is the consumer who decides which technology is desirable or not.

In most western societies, consumers are increasingly aware of health, environmental and animal welfare issues. Food safety and methods food production are part of their buying behaviour. However, price and production efficiency remain to be major contributors to sustainability of a livestock industry. Successful animal breeding programs need to find the right dose of technology that helps them to be competitive.

Breeding Strategies

Reproductive rate of breeding animals and uncertainty about true genetic merit of breeding animals make up the most important limiting factors in a breeding program. How many and which animals should be selected is determined by these factors. Investments in breeding programs are therefore often related to trait measurement and genetic evaluation, and to technology to increase reproductive rates.

Measurement Effort and Genetic Evaluation

The benefit of abundant and good measurement is that we may better be able to identify the genetically superior animals. This leads to more accurate selection and more genetic improvement.

Phenotypic measurements are turned into *Estimated Breeding value's (EBV's)*. Estimation of breeding value based on an animal's phenotype alone can already be quite accurate for high heritable traits. However, animals need to be compared across herds, and genetic and environmental influences have to be disentangled. To achieve this, more sophisticated statistical methods are used, leading to *Best Linear Unbiased Prediction* (BLUP) of breeding values. Besides allowing across herd comparisons, BLUP also uses all available information about an animals' breeding value, including data on related animals. Selection accuracy is strongly dependent on the degree of data recording, which requires a range of considerations related to cost and infrastructure.

In data recording, individual performances need to be related to animal identification. If BLUP is used to generate EBV's also an animal's pedigree needs to be known (in principle, for each animal only sire and dam). If pedigree is not recorded, breeding value can be assessed on own performance only, and is limited to sexes, which express the traits of interest. BLUP relies good structure of data (use of breeding animals across herds) and proper pedigree recording. If these prerequisites are in place, investment in BLUP methodology is usually highly cost efficient. Molecular genetic technology has rapidly developed in the past 2 decades. Genes have been found coding for factorial traits (such as many diseases). Many production traits are *quantitative traits* and a likely genetic model is here that genetic differences between animals are due to many genes. However, DNA technology has also provided genetic markers. Certain genetic markers can improve estimation of an animal's genetic potential as they are associated with regions that account for genetic variation. Genotyping animals for marker genotypes is therefore an investment with the aim to better assess true genetic merit of animals.

Reproductive Technology

Most of the main factors that determine genetic gain are directly influenced by the reproductive rate of the breeding animals. A higher reproductive rate leads to the need for a decreased number of breeding animals, therefore increasing the intensity of selection of these animals. If reproductive technology is possible, for example AI, the benefit could be expressed in terms of increased genetic rate of improvement, which in turn has a dollar component attached to it. More offspring per breeding animal allow also more accurate estimation of breeding value. Reproductive technology allows the intensive use of superior breeding stock.

An obvious consequence is possibly that the most popular breeding animals are overused, and the population could encounter inbreeding problems. Typically, as new technologies in animal breeding allow faster genetic change, long term issues such as inbreeding and maintenance of genetic variation become important. For that reason, selection tools in animals breeding have become somewhat more sophisticated in recent years. The impact of reproductive technologies on rates of genetic improvement and inbreeding will be discussed. Besides a direct effect on rate of genetic improvement, another important consequence from increasing reproductive rates is to disseminate superior genetic stock quickly. The influence of a superior breeding animal would be much higher if thousands of off spring could be born, rather than if the superiority is passed on through the production of sonsvia natural mating.

Another example is that of cloning. Cloning is not extremely important for increasing rate of genetic progress, but it could have a large impact by allowing many copies of the best individual to perform in commercial herds. As reproductive rates are basically multiplying factors in a breeding structure, any improvement in reproduction will justify higher investment in improvement of the best breeding stock.

Selection and Mating

The decision about which animals should be selected as parents for the next generation is mainly based on *assessment of breeding value* of individual animals. *Genetic evaluation* is central to animal improvement schemes. Selecting animals based on estimated breeding value maximizes the response to selection that can be achieved. However, there is one other criterion that is relevant when deciding which animals should have offspring.

This criteria is *common ancestry* of all selected parents. The coancestry of selected parents should stay below certain limits, since it is directly related to the build up of inbreeding. Coancestry among selected parents is determined by the average relationship among the selected parents as well as the number of parents selected. In this course we will more explicitly discuss selection strategies that maintain low levels of inbreeding. Decisions about which animals need to be mated are often seen in relation to dominance effects. Utilizing dominance variation is often not of primary importance for improvement of purebreds, but it can have more impact if breeding animals are selected from different breeds or lines, as heterotic effects between breeds can be utilized. When multiple traits are involved in the breeding objective, assortative mating could be useful, matching qualities in different parents for different. There is a good possibility that in the near future, planned mating will gain in importance, when effects of specific genotypes will be better understood. One could envisage certain genotypes with high growing potential to be combined with specific genes that have major effect on meat quality. Overview of animal breeding programmes avoid inbreeding in direct offspring as well as the rate of inbreeding in the population. However, the rate of inbreeding depends mainly on population size and number of parents selected. Methodology to optimize selection and mating decisions related to inbreeding will be discussed.

Structure of Breeding Programs

Most of the key decision factors mentioned earlier are related to the rate of genetic change that can be made. However, this could be genetic change in a small fraction of the national population (in nucleus or 'elite breeders'). Genetic superiority should be transferred as soon as possible to most of the commercial farms.

The structure of a breeding program is therefore relevant for two aspects of an improvement scheme:

Introduction to Marker Assisted Selection

Over the last two decades most livestock industries have successfully developed EBV's to allow identification of the best breeding animals. EBV's are best calculated using BLUP, meaning that they are based on pedigree and performance information of several traits from the individual animal and its relatives. BLUP EBV's are the most accurate criteria to identify genetically superior animals based on phenotypic performance recording. Although the idea of genetic selection is to improve the genes in our breeding animals, we actually never really observe those genes. Selection is based on the final effect of all genes working together, resulting in the performance traits that we observe on production animals. This strategy makes sense, since we select based on what we actually want to improve. However, animal performance is not only affected by genes, but also by other factors that we do not control. Selection for the best genes based on animal performance alone, can never reach perfect 100% accuracy.

A large progeny test comes close such a figure of perfect selection, but this is expensive for some traits (e.g. for traits related to meat quality), and we have to wait several years before the benefits from a progeny test have an effect. Efficient breeding programs are characterised by selecting animals at a young age, leading to a short generation intervals and faster genetic improvement per year. For selecting at younger ages, knowledge about the existence of potentially very good genes could be very helpful.

Quantitative genetics uses phenotypic information to help identify animals with good genes. *Extension to use information from molecular* genetics techniques aim to locate and exploit gene loci which have a major effect on quantitative traits (hence QTL-Quantitative Trait Loci). The idea behind marker assisted selection is that there may be genes with significant effects that may be targeted specifically in selection. Most traits of economic importance are quantitative traits that most likely are controlled by a fairly large number of genes. However, some of these genes might have a larger effect. Such genes can be called major genes located at QTL.

In practice, we rarely know the genotype at actual QTL, as the exact gene location (mutation) is often unknown. Currently there are few examples where QTL effects can be directly determined, but knowledge in this area is rapidly developing. Most QTL known today can only be targeted by *genetic markers*. Genetic markers are "landmarks' at the genome that can be chosen for their proximity to QTL. We cannot actually observe inheritance at the QTL itself, but we observe inheritance at the marker, which is close to the QTL. When making selection decisions based on marker genotypes, it is important to know what information can be inferred from the marker genotypes.

We can identify the marker genotype (Mm) but not the QTL genotype (Qq). The last is really what we want to know because of its effect on economically important traits. Let the Q allele have a positive effect, therefore being the preferred allele. In the example, the M marker allele is linked to the Q in the sire. Progeny that receive the M allele from the sire, have a high chance of having also received the Q allele, and are therefore the preferred candidates in selection.

Mixed Models in Animal Breeding: Where to Now?

Over the past 60 years, mixed models have underpinned huge gains in plant and animal production through genetic improvement. Charles Henderson (1912-1989) established mixed models for estimating breeding values (BLUP) using the popularly called Henderson's Mixed Model and provided early methods (Henderson's Methods I, II and III) for estimating variance parameters. Robin Thompson then published the widely acclaimed REML method for variance component estimation in 1971. These two innovators, along with the development of computing power, have spawned national and international breeding programs in almost all animal species used for human food and fibre.

Our ability to generate data is outstripping our ability to analyse data and this will lead to mixed models playing new roles in genetic estimation. The focus is changing from simply describing the relationship between variables through a correlation, to modelling the relationship based on knowledge of the Genome.

Selective breeding goes back at least to Jacob (1800 BC, Genesis 30) who selected the fitter rams for his own flock. Traditional breeding has largely relied on visual assessment with many such classers having considerable skill in recognising genetic potential with respect to their objective, whether breeding war horses, dogs or pigeons. What characterises modern breeding though is the extensive use of objective measurement and adjustment for environmental effects.

The digital age has seen a rapid increase in the number of traits included in a breeding objective or selection criterion, as well as use of data on relatives to improve the separation of genetic from environmental differences. Charles Henderson (1912-1989) *et al.* (1949, 1959) developed and popularised the mixed model equations which underpin the BLUP estimation of breeding values. His development of these equations included use of the additive genetic relationship matrix, showing how it accommodates selection as well as their primary role of adjusting for nuisance environmental effect

However, the mixed model equations used for evaluation assume knowledge of variance parameters. Henderson (1953) defined the main methods used to estimate these until Robin Thompson (Patterson and Thompson 1971) presented the Residual Maximum Likelihood (REML) method. Karin Meyer and Dorothy Robinson produced software to implement REML methods (in animal breeding and However analysis was difficult until Robin presented the Average Information method underpinning ASReml (1997, 2002, 2006, 2009) which become generally available in 1997.

The promise of the genomic revolution is that we may be able to select directly for specific combinations of genes based on reading an individual's genetic code and having good information on the phenotypic and pleiotropic effects of genes/alleles.

Mixed Model Equations and Blup

The linear mixed model is written as where X is the design matrix for fixed effects, τ, Z is the design matrix for random effects, u, y is the vector of phenotypic measurements and e is the vector of model residuals. The mixed model equations (MME) are conveniently represented in matrix form by where var. Given R and G, the solution for the fixed effects given by the mixed model equations is the same as given by solving where. The solutions for the random effects are the Best Linear Unbiased Predictors of those effects and as such are ideal for selecting breeding stock.

The power of this system lies in the structure that can be incorporated into X, Z, R and G. It is is not unusual for u to include sub-vectors for various traits and various 'strata' such as direct genetic, maternal genetic, maternal environment, dominance and nuisance blocking effects. This can lead to a fairly complex structure to G involving relationship matrices and variance matrices of various sorts. The main advantage of the mixed model equations is that the left hand side matrix is typically fairly sparse so that large systems of equations can be solved quite efficiently. This arises because matrices X and Z, and inverses of R and G are typically sparse.

Residual Maximum Likelihood

Without going into the detail, suffice to say that if we assume u and e (and therefore y) are normally distributed (given R and G), we obtain an expression involving y, X, Z, R and G which is called the likelihood. This expression can be partitioned into two parts; one providing information on τ conditional on G and R leading to the mixed model equations, the other providing information on R and G conditional on τ. Residual Maximum Likelihood seeks to find the parameter values for R and G that are most likely because they maximise this second part (rather than the whole likelihood). This maximisation exercise though was not trivial when R and G involved more than a few parameters and the problem was large. Consequently, REML estimation was restricted in application to small problems or well structured standard animal breeding models until Thompson presented the Average Information procedure which is also centred around the mixed model equations. The implementation in ASReml exploits the sparsity of the mixed model equations though judicious ordering of the equations, avoiding the need to obtain the complete inverse of the left hand side matrix. Now REML can be applied to large problems (with several hundred variance parameters).

Where to Now?

One thing programming has taught me is that no matter how big you allow, someone will want bigger. While computing technology has helped with the more traits, more records issue of modern animal breeding based on BLUP technology, we are now faced with genome level data of a higher magnitude and methodologies which do not have the statistical and mathematical rigor that supports conventional quantitative genetics. Three problem areas come to mind. The first is the well established variance estimation problem (Hill and Thompson 1978) that when estimating a variance matrix, the probability that the maximum value of the REML likelihood occurs outside the imposed parameter space increases with the matrix size. The second is the application of mixed models to genomic data. The third is how to effectively combine specific genomic data into the BLUP evaluation process.

Structured Variance models. The more traits involved in a REML analysis, the more likely there will be difficulties with the estimation of all the variances and co-variances involved. ASReml will estimate a negative definite matrix if permitted, or attempt to estimate a positive definite matrix which is almost singular. But this raises the issue of whether a reduced parameterization within the parameter space will be preferable. It is not uncommon to find that a matrix can be reduced by use of principal components to a more parsimonious form. That is, the first 1, 2 or 3 principal components will contain the big bulk of the information contained in the matrix. The remaining variation is noise and is often associated with negative eigen values. Therefore it makes sense to estimate the matrix based on some underlying structure. Three structures are common in ASReml. For variates that have no intrinsic ordering, the principal component/factor analytic models allow more parsimonious modelling. For measurements repeated at irregular intervals, the random regression models are often applied but these may produce unreasonable estimates at the ends of the time range.

For regular repeated traits, for example weights at successive ages, the expected structure is an autoregressive one for which the Antedependence (Generalised auto regressive) models apply. Jaffrzic et al. (2002) has extended the Antedependence model to a Structured antedependence where a model is imposed on the regression and innovation parameters. Meyer and Kirkpatrick (2009) have investigated a reduced parameterization based on assuming common eigen vectors across strata which is another proposal within this framework. To my mind, this leads to a general area of writing models for the variance parameters, and is the next logical step when it comes to fitting models with hundreds of variance parameters. The question will always be whether a reduced parameterization has adequately captured the real variation without imposing a structure unsupported by the data. Mixed models for genomic data. There is a huge literature on analysing the huge amount of genomic data that is being presented and little consensus on the best approach. One issue is the diversity of kinds of data available and the other is the sheer volume of data and the knowledge that meaningful/useful variation is present in only a small proportion of it. The issue here is then to separate signal from noise. I believe mixed models could have a bigger role here because signal will represent a covariance (or inflated variance) over the noise (base variance). Mixed models have been successfully used to adjust for spatial variation in genomic slides. They have been used to locate QTL in back-cross/F2 experiments and in association studies where there are often more 'markers' than experimental units. Thomson et al. (2009) use mixed models as part of their procedure to combined cattle and sheep genomic data to look for differentially expressed genes. The new outlier method in ASReml 3 may help in this regard. Incorporating genomic markers in BLUP evaluation. Scientists are an optimistic group when it comes to incorporating genetic markers into BLUP evaluation. I suspect there is a lot of detailed work required before this becomes standard procedure across the industries.

Discussion

Linear Mixed Models have underpinned a revolution in livestock breeding in the last 50 years and despite the huge investment in genomic research and Bayesian methods, there remains a continuing major role for them in the foreseeable future. However, the general model needs adaption for the specifics of each particular species and application. By this I mean, identification of the principle sources of variation, whether they should be accommodated as fixed or random effects, appropriate variance structures and extending the analyses to larger populations and with more traits. While a bivariate analysis is now readily performed, larger multivariate analyses for the estimation of positive definite variance matrices are often difficult requiring use of structured matrices and raising the issue of whether the structure is adequate. There will undoubtedly be further developments in this area. The literature on analysis of genomic data reports a wide range of methods as people have hurried to analyse their large amounts of newly acquired data. Some of these analyses have demonstrated the utility of mixed models in this area, but have also shown up limitations due to the amount and structure of the new data. This also will need more attention.

4

Reproductive Methods in Low Input Animal Breeding

Generally we should declare that ethics cannot be described by exact parameters but it is the conclusion of some principals and long lasting social rules directing human behaviour. It is a similar phenomenon in animal breeding and mostly in that type of breeding which has historical traditions and emotional relation to country people. Since ethical issues have been getting rather "artificial" explanation by the media recently, people in the cities and unfortunately sometimes in the countryside many times do not have proper information on this field.

Nevertheless reproductive researchers also do not have opinion overcoming all others usually they have a modest and realistic behaviour in their work aiming to improve the farmers' results in their daily practice.

Ethical aspects of reproductive methods in low input breeding (LIB) need a rather complex approach determined by:

1. Human factors – rural local and regional healthy food supply, rural development i.e. reducing unemployment, maintaining agriculture traditions, support rural tourism and environment protection.

2. Animal factors – gene conservation by all possible and necessary methods as well as animal welfare in the balanced frame of traditions and actual demands according to the forthcoming challenges of the rapidly changing world.

Easy to see that even by the applied reproductive management we cannot respond all criteria Sabove thus we should select the priorities with regard on main purpose of the farm in case. Before discussing the different techniques it should be declared that reproductive methods even in LIB should not miss the well trained experts having relevant knowledge and experience. Animal health conditions, individual registration of animals, exact documentation of mating or artificial insemination (AI) are required at the same manner as in intensive farming. It means that professionals involved in LIB should be familiar not only with the traditional methods but with the newest results of innovation, as well.

We try to give over some of our experiences regarding on reproductive techniques in LIB which has a special importance also in Hungary and we were conducting international research projects connected to it. Animal breeds in LIB are mostly indigenous and sometimes endangered ones adapted to the climate of Carpathian Basin very well during recent centuries.

These animals are incorporating our traditions, representing our agricultural national value and capable for playing a key role in rural development, rural tourism and first of all unique processed products can be produced of their carcass i.e. Hungarian winter salami, sausage, bacon etc.

Mangalica Pig

Mangalica pigs, more precisely Blond, Red and Swallow Belly Mangalica were always bred in small and large scale farms in the past and it is the same in the present. Apart from their own production large companies integrated small farmers' activity and organized the breeding and trading. The Mangalica Breeding Association is giving guidance for the members and supply boars or semen for them.

Reproductive work is done by natural mating and AI both in small and large farms. In some blood strains the low number of animals evidently needs natural mating for in vivo gene preservation. In some production units AI is a daily practice.

We should underline the necessity of AI in any production units. It is not so much costly as keeping minimally 5 times more boars but highly dedicated work is indispensable. Generally the AI is significantly different and probably more complicated in native breeds than in their modern counterparts. It is true in terms of male as well as female reproductive physiology. Semen deep freezing is a connecting wing of LIB when involved farmers and companies are cooperating with research units and they develop together the most important tool for in vitro gene conservation. Although, boar semen cryopreservation is not clearly solved, our group has some really promising results (average 50% post-thaw motility).

National parks have an emphasized duty to demonstrate our agriculture traditions and they should keep indigenous domestic animal breeds e.g. Mangalica pigs in pure bred population among old, typical LIB circumstances.

It can be concluded that LIB in Hungarian Mangalica sector found its proper position and small farmers can take part of local and regional food supply by high value processed pork products. (Unfortunately recently upcoming high feed prices are very harmful for Mangalica breeders, too.)

Several successful attempts are running in Europe using also commercial breeds in organic farming for producing "BIO" labelled meat and processed products.

Racka Sheep

Racka sheep has history of several hundred years. The Hungarian and other nomad tribes had similar type of sheep already on the Middle Asian steppes and we can find nearly identical breeds there even now.

It has a diminished population in Hungary with two colour types i.e. the White and Black Racka. National parks and enthusiastic sheep breeders keep, also some village hotels keep them in small units. Although some attempts have been done to establish a new market role for it lower meat yield % (in EUROPE classification) undermined the efforts till now.

Definitely an other evaluation system would be desirable for the ancient breeds. Anyway, Racka is excellently suitable for LIB in continental climate of the Carpathian Basin thus it could contribute rural development programs in remote areas.

Actually eminent purpose of reproductive management declared by the Hungarian Sheep Breeders' Association is the preservation of this breed. Almost everywhere ewes are mated naturally by selected rams.

LIB has a special significance in modern animal breeding. Both in rural tourism and rural development it can find its proper place but farmers involved should clearly know their role inthe sector. If they choose small scale farming, they must use the relevant techniques e.g. reproductive methods and if they decide to take part of larger production amount they should adapt modern and more effective system. Innovation is always necessary either by taking part in it or by collecting the available achievements. In our consideration reproductive methods in LIB does not mean closed eyes and ears but to be opened for the new results enabling the farmers to improve their dedicated work.

Genetic Engineering Applications in Animal Breeding

Genetic engineering is the name of a group of techniques used for direct genetic modification of organisms or population of organisms using recombination of DNA. These procedures are of use to identify, replicate, modify and transfer the genetic material of cells, tissues or complete organisms. Most techniques are related to the direct manipulation of DNA oriented to the expression of particular genes. In a broader sense, genetic engineering involves the incorporation of DNA markers for selection (marker-assisted selection, MAS), to increase the efficiency of the so called 'traditional' methods of breeding based on phenotypic information. The most accepted purpose of genetic engineering is focused on the direct manipulation of DNA sequences These techniques involve the capacity to isolate, cut and transfer specific DNA pieces, corresponding to specific genes.

The mammalian genome has a larger size and has a more complex organization than in viruses, bacteria and plants. Consequently, genetic modification of animals, using molecular genetics and recombinant DNA technology is more difficult and costly than in simpler organisms. In mammals, techniques for reproductive manipulation of gametes and embryos such as obtaining of a complete new organism from adult differentiated cells (cloning), and procedures for artificial reproduction such as in vitro fertilization, embryo transfer and artificial insemination, are frequently an important part of these processes.

Current research in genetic engineering of animals is oriented toward a variety of possible medical, pharmaceutical and agricultural applications. Also, there is an interest to increase basic knowledge about mammalian genetics and physiology, including complex traits controlled by many genes such as many human and animal diseases. The interest in genetic engineering of mammalian cells is based in the idea of, for example, use gene therapy to cure genetic diseases such as cystic fibrosis by replacing the damaged copies of the gene by normal ones in foetuses or infants (gene therapy). Genetically engineered animals such as the 'knockout mouse', in which one specific gene is 'turned off', are used to model genetic diseases in humans and to discover the function of specific sites of the genome (Majzoub and Muglia, 1996). Genetically modified animals such as pigs will probably be used to produce organs for transplant to humans (xenotransplantation). Other applications include production of specific therapeutic human proteins such as insulin in the mammary gland of genetically modified milking animals like goats (transgenic animals, bioreactors). These techniques may be used to increase disease resistance and productivity in agriculturally important animals by increasing the frequency of the desired alleles in the populations used in food production. This can be accomplished by transferring alleles or allele combinations, over expressing or eliminating the expression of particular genes (use of genetic engineering in animal breeding). In addition, these techniques open the possibility of using artificially modified genes to increase the biological efficiency of proteins (Kinghorn, 2003).

The objective of this paper is to review some advances on genetic engineering applications in animal breeding, including a description of the methods, some applications and ethical issues. Here I made emphasis in both the search and use of genomic information for selecting animals and to transfer and use their genes in commercial populations via marker-assisted selection (MAS) or transgenesis.

This review focuses mainly in the methodology to apply genetic engineering directly to animals for genetic improvement.

Several important biotechnological applications such as the production of recombinant proteins in bioreactors (Houdebine, 2002), disease diagnostic (McKeever and Rege, 1999), feedstuff processing (Bonneau and Laarveld, 1999) and production of vaccines (Eloit, 1998), proteins, stem cells, tissues and monoclonal antibodies for use in therapeutics are not included here. The impact of reproductive technologies on animal breeding, not directly related with gene transfer, are reviewed elsewhere. The possible role for cloning adult animals in breeding is also discussed

Use of Genomic Information in Animal Improvement

The use of genomic information (sequences or DNA marker polymorphisms) for the genetic improvement and selection of animals requires the knowledge of the effect of physically mapped genes with effects on economically important traits or quantitative trait loci (QTL). This information is also required in order to effectively use transgenesis and MAS for genetic improvement. In MAS, the genomic information is combined with the classical performance records and genealogical information to increase selection accuracy, performing selection earlier in life and reducing costs.

The traits on which the application of marker-assisted selection can be more effective, are those that are expressed late in the life of the animal, have low heritability, are sex-limited, are expensive to measure or are controlled by a few genes. Examples are longevity, carcass traits in meat producing animals, and diseases or defects of simple inheritance.

Expected increments in selection response from MAS for a single complex trait, using known QTL genotypes plus linear model predictions (BLUP), compared to selection on BLUP alone, ranges from-0.7 to 64 percent. In practice, results will depend on many parameters which are likely to be very different for each trait combination and population. The statistical properties of genetic evaluations (predictions) of animals for quantitative traits obtained through mixed model methodology using phenotypic records and genealogical information as inputs are known as BLUP. Best-means minimum variance of prediction, Linear-because predictions are linear functions of observations, Unbiased-means that the expected value of predictors obtained with linear model have an expected value equal to the expected value of the mean of the breeding values, conditional to data, and Prediction-because involves prediction of random breeding values).

Most experiments on QTL detection in animals allow only the estimation of wide chromosomal regions (practical maximum resolution is of about 1 cM, but usual resolution is about 30 cM) that harbour a QTL in a 'statistical sense', estimated from the effects of some marker haplotypes on quantitative traits (de Koning et al. 2003). Thus, further confirmation is required in order to assure the use of the causative gene. Identification of the causative gene has proven to be difficult.

The process to identify the gene responsible for the effect is known either as 'fine mapping' studies (targeting mapping smaller genomic regions) or 'candidate gene' studies (targeting individual genes based on their probable function) (Lynch and Walsh, 1998). In practice, MAS is useful to select genes with effects well identified and precisely located in the genome such as those controlling monogenic recessive diseases such as the pig stress syndrome gene. However, for most recessive alleles with lethal or semi-lethal effects, natural selection will maintain their frequencies very low (Hartl and Clark, 1997) making MAS unnecessary.

Unless the additive and non-additive effects for most genes involved in the phenotypic expression of complex, economically important traits are determined, MAS should be regarded just as a tool to increase the rates of genetic gains and not a method to fully open the 'black box' of the genetic control of complex traits, that would render phenotypic selection 'obsolete'. Therefore, the perspectives on the optimum use of DNA marker information in the framework of a genetic program is still a matter of debate.

Quantitative trait loci experiments using crosses between breeds or lines with extreme genotypes for a trait, increases the power of detecting QTLs for that trait, compared to within-family designs. These across population's polymorphisms are not necessarily useful to perform MAS for within-population selection. The favourable allele could be fixed in parental populations and crosses may be commercially irrelevant. Wide genome scans for positioning a QTL using crosses or within-family experiments, are only the initial phase of the search for a true mayor gene involved in a complex trait (de Koning et al. 2003). Another source of complexity for detection and use of QTL for selection is genetic heterogeneity, where DNA mutations in several sites produce the same phenotype. Major single gene effects can be sometimes compensated in the organism using alternative metabolic pathways (McAfee, 2003).

Problems related to false positive detection of candidate genes are also common. Using crosses between two pig breeds, a polymorphism on the estrogen receptor locus (ESR) was associated to litter size in pigs with 1.5 piglet advantage for homozygous sows for the beneficial allele, and where followed by immediate recommendations for commercial use and patenting (Rotschild et al. 1996). Further research however did not confirm the effect.

Different phases of linkage between the markers and the QTL could explain the fact that the effect of the ESR locus varied widely between populations (Gibson et al. 2002). Thus, very probably, despite the ESR gene is probably a plausible 'candidate' from their inferred physiological functions (Rotschild et al. 1996), the gene involved seems to be another one, still unknown, or the effect initially observed was the product of several, interacting genes (epistasis).

Main problems related to the use of molecular genetics in the improvement of agricultural populations are:

1. Direct use of a discovered QTL effect for selection across families is not possible.

2. By the time the information about the inferred genotypes is known, frequently the animals involved in the study are not available as candidates for selection, because they will be too old.

3. Advantage from within-family selection for a QTL bracketed by markers over BLUP or phenotypic selection alone is frequently low and the methodology to exploit this information for selection is complex and relatively inefficient.

4. There are statistical estimation errors, causing both false positive and false negative effects, particularly when the effect of the QTL is small.

5. There is a lack of consistency of the effect of the same QTL between studies, caused by QTL by genetic background (epistasis) of QTL by environment interactions.

6. The net economic effect of the QTL may be lower than the effect on single traits, because unfavourable effects on other traits.

7. Selection using QTL is more complex than phenotypic selection alone. QTL information (whether the information on the QTL is direct or indirect), adds to the list of traits used as selection criteria. Issues such as reduction of selection intensities and relative emphasis given to each trait, make optimal selection more difficult, with a need for adequate relative weights for the QTL, and the polygenic portions of the genetic variation for each trait at each generation (year).

8. Short-term gains due to MAS may be at the expense of medium to long-term polygenic responses for important traits.

Even with an unambiguous knowledge for the allele effects of a mayor gene on a complex trait, expected advantages from optimum use of genotyping alleles for a QTL for a multi-generation selection horizon is not always high. The polymorphism for the αs1-casein in goats has a strong effect on protein content and total protein output. The difference between homozygous for the highest and lowest effects for milk protein is approximately three phenotypic standard deviations for milk protein content. Favourable alleles have frequencies lower to 0.5 in populations undergoing selection, making a very favourable case for potential gains in protein content and production from MAS using this polymorphism.

Simulation studies by Larzul et al. (1997), Fournet et al. (1997) and by Manfredi et al. (1998) indicated that when an efficient 'conventional' progeny testing selection program is underway for increased protein production, the advantages from MAS are low to moderate. Maximum possible increase on total genetic gain for protein yield was 26%. Dekkers and Hospital (2002) emphasized the overlap that exists between marker and phenotypic information for the improvement of a multi-trait goal over several generations, using MAS. A very optimistic prospect from use of MAS as well as other biotechnologies is very common in popular commercial and non-refereed publications, based on approaches based on exploiting single gene effects, without consideration to polygenic effects, economic values or time for fixation.

Research shows that the real situations are far more difficult for complex traits. These traits are controlled by several genes and environmental effects. Dekkers (2004) made a survey on the status of application of MAS in actual animal breeding programs for complex traits. He concluded that initial expectations for the use of MAS were high, but the current attitude is one of cautious optimism, with a need for careful examination of alternative selection strategies, business goals and integration of molecular and other technologies. Pollak (2005) made a detailed survey on the application of DNA technology for beef cattle improvement in USA. He concluded that current contribution of the new DNA technologies for beef cattle breeding is marginal, because they are encountering logistics and mechanical issues. For genomics technologies to impact fully on the beef industry, a higher level of sophistication of the genetic tests will be needed. Tests based on the genes themselves, rather than DNA markers associated with genes, will be required.

It is theoretically possible to predict accurately the breeding values of animals using many markers (Meuwissen et al. 2001). From this knowledge, it is possible to develop a model for *in vitro* genetic improvement of animals. This is known as velogenetics. The model involves *in vitro* selection of cells containing the desired genes the use of totipotent embryonic stem cells (ES). The procedure uses transfection of the desired genes, selection in vitro of the cells, and nuclear transfer of the desired genotypes into receptor oocytes. This approach is supposed to increase the rate of genetic improvement by obtaining many generations in a short time by avoiding rearing, reproduction and selection of 'real animals'. Selection on the basis of genomic information only, such in this *in vitro* system, even with major genes with known effects well localized, may be dangerous, because in these artificial populations, unlike in real populations, natural selection would not be allowed to act at each generation on fitness traits under real, perhaps changing, environmental conditions. Changes on economically important traits will not be evaluated directly (Dekkers and Hospital, 2002). This may potentially reduce the responses on selected traits because of genotype x environment interactions (Montaldo, 2001). This is because selection is performed in artificial conditions that may deteriorate the fitness of the population and economic response.

Using MAS for improving health in animals by reducing disease prevalence (increasing disease resistance) or increasing resilience (the ability to withstand the disease without harmful effects), for infectious or parasitic diseases has been difficult. In most cases, excepting some rare examples such as Scrapie in sheep, complete resistance could not be obtained with the manipulation of a small number of genes. For most diseases, single-gene approaches are expected to have only a partial contribution. Gene interactions are common (Kuhnlein et al. 2003).

For many diseases, heritabilities are often low. That indicates the existence of many environmental factors affecting both the probability of infection and the response of the host. In spite of responses attained using conventional selection for some traits that are used as indicators of disease, the result is not well known. The existence of contradictory results regarding associations between production and disease resistance, the complexities of immune and resistance mechanisms and the interaction with other methods of control such as vaccination, sanitation, management and chemotherapy, makes the whole issue of selecting for disease resistance more difficult, in principle, that selecting for production traits. Moreover, we know that heritable resistance or resilience to more virulent form of pathogens would be increased by natural selection. As heritabilities for survival are generally low, we know that the genetic control of disease may be very complex, making difficult to change the outcome by manipulating single genes.

There is one published result on a successful MAS selection program to reduce the prevalence of dermatophilosis, a tropical infectious disease in Zebu cattle (Maillard et al. 2003). Maillard et al. (2003) argue to have obtained a sharp reduction in clinical prevalence of the disease from 0.76 to 0.02 in a period of five years by selecting against only two type II BoLA alleles associated with a high susceptibility of the disease. The authors explained the observed change resulting from selection performed in an unknown number of animals of each sex in 1996. However, a complete description of the changes in allele frequencies and genotypes from the moment of selection and their association with the evolution of prevalence by sex is not given. Considering the possibility of environmental changes and the presence of natural selection, in the absence of a control group, it is difficult to know if the observed change is the sole result of the mechanisms invoked by the authors through MAS.

We cannot at this moment forecast precisely the future of MAS in animal selection, but it is premature to conclude that methods based on phenotypic information will be replaced by methods based solely on genomic data. An integration of both types of data with the use of more sophisticated statistical models is needed. It is far from sure that total replacement of phenotypic information with gene-by-gene information, as selection criterion is possible or even desirable in the future.

Other very important applications of genetic markers in animal improvement include the optimization of mating strategies for non-additive genetic effects (estimation and managing of inbreeding and heterosis), parentage determination, genetic characterization of diverse animal breeds and populations using studies of between and within population (breeds) diversity (Oldenbroek, 1999) and marker-assisted introgression of particular alleles.

Cloning Adult Mammals

Cloning an animal is the production of a genetically identical individual, by transferring the nucleus of differentiated adult cells into an oocyte from which the nucleus has been removed. This is known as Nuclear Transfer and is how the Dolly sheep was produced.

Since the publication of the original paper on cloning (Wilmut et al. 1997), there are several other reports on adult cloned animals involving mice, cattle, cats, goats, pigs, sheep and rabbits involving the same, and other cloning techniques.

Cloning Methods

In the case of Dolly, mammary gland cells in culture from a 6-year old donor ewe, where subjected to a reduction in the concentration of serum and thus obliged to enter in a quiescent state of the cell cycle (G0). Nuclear transfers to enucleated oocytes, was followed by electrical pulses for fusion of the donor cell nucleus and oocyte membranes and activate division (Wilmut et al. 1997).

Problems

Currently there is no doubts regarding the genetic similarity of the donor and the clone in the case of Dolly, however, besides low success rates (Edwards et al. 2003), several health problems related to the technique have been described (Samiec and Skrzyszowska, 2005). Normal development of an embryo is dependent on the methylation state of the DNA contributed by the sperm and egg and on the appropriate reconfiguration of the chromatin structure after fertilization. Somatic cells have very different chromatin structure to sperm and 'reprogramming' of the transferred nuclei must occur within a few hours of activation of reconstructed embryos. Incomplete or inappropriate reprogramming will lead to de-regulation of gene expression and failure of the embryo or foetus to develop normally or to non-fatal developmental abnormalities in those that survive. These facts indicate that there is a need for studies to determine further biological consequences of cloning. Cloning has important potential applications in gene transfer procedures.

Use of Cloning in Animal Breeding

Use of cloning in animal genetic improvement may increase the rates of selection progress in certain cases, particularly in situations where artificial insemination is not possible, such as in pastoral systems with ruminants. Currently, high costs of cloning are one of the main factors limiting their use as a technique in practical animal breeding. Clonal groups, however more uniform than full sibs, will have all differences caused by the environmental fraction of variation for measured traits, which is usually more than 50% of total variation. Selection among many cloned germlines allows the use of the non-additive genetic effects. These effects are not exploited when traditional selection methods involving sexual reproduction are used in animal improvement (Visscher et al. 2000), but most of the observed genetic variation between animals is additive (Van Vleck, 1999). Advantages in terms of additional genetic progress however, seems to be only marginal from clone evaluation in selection nucleus herds (Ruane et al. 1997). Production based on clones of the best animals of the population, may allow for a one time large 'jump' in breeding value, so the commercial animals might be very close to those in the nucleus. However, further genetic improvement must be based in the continued use of the genetic variation by selection programs.

Transgenic Animals

Transgenesis is a procedure in which a gene or part of a gene from one individual is incorporated in the genome of another one. Transgenic animals have any of these genetic modifications with potential use in studying mechanisms of gene function, changing attributes of the animal in order to synthesize proteins of high value, create models for human disease or to improve productivity or disease resistance in animals. In the early 80´s, several research groups reported success in gene transfer and the development of transgenic mice. The definition of transgenic animal has been extended to include animals that result from the molecular manipulation of endogenous genomic DNA, including all techniques from DNA microinjection to embryonic stem (ES) cell transfer and 'knockout' mouse production (Cameron et al. 1994). Since the early 1980s, the production of transgenic mice by microinjection of DNA into the pronucleus of zygotes has been the most productive and widely used technique. Using transgenic technology in the mouse, such as antisense RNA encoding transgenesis, it is now possible to add a new gene to the genome, increase the level of expression or change the tissue specificity of expression of a gene, and decrease the level of synthesis of a specific protein. Removal or alteration of an existing gene via homologous recombination required the use of ES cells and was limited to the mouse until the advent of nuclear transfer cloning procedures.

Transgenic Methods

Microinjection of DNA and now nuclear transfer, are two methods used to produce transgenic livestock successfully. The steps in the development of transgenic models are relatively straightforward. Once a specific fusion gene containing a promoter and the gene to be expressed has been cloned and characterized, sufficient quantities are isolated, purified and tested in cell culture if possible and readied for preliminary mammalian gene transfer experiments. In contrast with nuclear transfer studies, DNA microinjection experiments were first performed in the mouse (Izquierdo, 2001). While the transgenic mouse model will not always identify likely phenotypic expression patterns in domestic animals, there have not been a single construct that would function in a pig when there was no evidence of transgene expression in mice. Preliminary experimentation in mice has been a crucial component of any gene transfer experiment in domestic animals (Kerr and Wellnitz, 2003). While nuclear transfer might be considered inefficient in its current form, major advances in experimental protocols, can be anticipated. The added possibility of gene targeting through nuclear transplantation opens up a host of applications, particularly with regard to the use of transgenic animals to produce human pharmaceuticals. The only major technological advance since the initial production of transgenic farm animals has been the development of methods for the *in vitro* maturation of oocytes (IVM), *in vitro* fertilization (IVF) and subsequent culture of injected embryos prior to transfer to recipient females. Another highly efficient technique for transgenesis has been recently developed based on the use of lentiviral vectors to transform cow and pig oocytes. These vectors are more efficient than microinjection in terms of transformation and expression rates. One limitation is that the size of the transgene and the internal promoter has to be less than 8.5 kb in size.

Transgenesis in the Improvement of Production traits

The technology of transgenesis is potentially useful to modify characters of economic importance in a rapid and precise way. Contrary to the 'classical' selection programs, it is necessary a knowledge of the genes that control these characters and their regulation.

Following is a brief discussion of experiences with transgenesis to alter economically important traits in livestock.

Growth and Meat Traits

In most of the earlier work in domestic species (pig, sheep, rabbit) growth hormone was enhanced by the metallotionein promoter to control its expression. Subsequent efforts to genetically alter growth rates and patterns have included production of transgenic swine and cattle expressing a foreign c-ski oncogene, which targets skeletal muscle, and studies of growth in lines of mice and sheep that separately express transgenes encoding growth hormone-releasing factor (GRF) or insulin-like growth factor I (IGF-I). Transgenic pigs and sheep with high levels of serum growth hormone were obtained, but an increment of its rate of growth was not observed, and only in some lines average daily gain increased with the supplement of the diet with high levels of protein. The highest effects were observed in the reduction of body fat. A large number of different serious pathologies and a severe reduction in reproductive capacity were described in these animals (Murray et al. 1999). In a report about two studies with pigs (Neimann, 1998), there is evidence for the use of transgenesis allowing to important reductions in body fat and increased diameter of muscle fiber by increased IGF-I levels and growth hormone without serious pathological side effects. Australian regulations avoided the commercial release of these animals.

Frequently the used promoters have not allowed an efficient control of the expression of the transgene. It was assessed that it is necessary to develop more complex constructions that activate or repress the expression of the transgene more precisely. Adams et al. (2002) found inconsistent results regarding the effect of a growth hormone construct in sheep on growth and meat quality.

Recently, a spectacular transformation was obtained by insertion of a plant gene in pigs. Saeki et al. (2004) generated transgenic pigs that carried the fatty acid desaturation 2 gene for a 12 fatty acid desaturase from spinach. Levels of linoleic acid (18:2n-6) in adipocytes that had differentiated in vitro from cells derived from the transgenic pigs were 10 times higher than those from wild-type pigs. In addition, the white adipose tissue of transgenic pigs contained 20% more linoleic acid (18:2n-6) than that of wild-type pigs. These results demonstrate the functional expression of a plant gene for a fatty acid desaturase in mammals, opening up the possibility of modifying the fatty acid composition of products from domestic animals by transgenic technology.

Wool Production

The objectives are to improve production of sheep wool and to modify the properties of the fiber. Because cystein seems to be the limiting amino acid for wool synthesis, the first approach was to increase its production through transfer of cystein biosynthesis from bacterial genes to sheep genome (Murray et al. 1999). This approach did not achieve the efficient expression of these enzymes in the rumen of transgenic sheep.

Milk Composition

Milk proteins are coded by unique copy genes that can be altered to modify milk composition and properties. Among the different applications of milk modification in transgenic animals, the following can be highlighted:

1. To modify bovine milk to make it more appropriate to the consumption of infants. Human milk lacks β-lactoglobulin, has a higher relationship of serum proteins to caseins, and has a higher content in lactoferrin and lysozyme when compared to bovine milk. Lactoferrin is responsible for the iron transport and inhibits the bacterial growth. To introduce the human lactoferrin into the bovine milk, transgenic cows have been obtained. The elimination of the β-lactoglobulin in the cow milk would be another interesting objective because is one of the major allergens of cow's milk.

2. To reduce the content of lactose in the milk to allow their consumption to people with intolerance to lactose. It is considered that 70% of the world population is lacking theintestinal lactase, the enzyme required to digest the lactose. The reduction in lactose may be obtained by expressing β-galactosidase in the milk or diminishing the content of α-lactalbumin. Transgenic mouse with inactivated α-lactalbumin gene produce milk without lactose. However, a serious practical drawback of this method is that this milk is very viscous and it is not secreted to the exterior of the mammary gland, due to the importance of the lactose in the osmoregulation of the milk (Stinnakre et al. 1994).

3. To alter the content of caseins of the milk to increase their nutritive value, cheese yield and processing properties. Research has intended to increase the number of copies of the gene of the κ-casein, to reduce the size of the micelles and modificating the κ-casein to make it more susceptible to the digestion with chymosin. This has only been done using the mouse as a model (Gutiérrez-Adan et al. 1996). Brophy et al.(2003) engineered female bovine foetal fibroblasts to express additional copies of transgenes encoding two types of casein: bovine β-casein and κ-casein. The modified cell lines of fibroblasts were used to

create eleven cloned calves. Milk from the cloned animals was enriched for β-and κ-casein, resulting in a 30% increase in the total milk casein or a 13% increase in total milk protein, demonstrating the potential of this technology to make modified milk.

4. To express antibacterial substances in the milk, such as proteases to increase mastitis resistance. The objective is to alter the concentrations of antibacterial proteins such as lyzozyme or transferrin in the milk.

Future Perspectives of Transgenesis

The techniques for obtaining transgenic animals in species of agricultural interest are still inefficient. Some approaches that may overcome this problem are based on cloning strategies. Using these techniques it is feasible to reduce to less than 50% the number of embryo receptor females, which is one of the most important economic limiting factor in domestic species. It would also facilitate the further proliferation of transgenic animals. Recent results relate these techniques with still low success rates (Edwards et al. 2003), high rates of perinatal mortality and variable transgenic expression that requires to be evaluated before generalizing their application.

Considerable effort and time is required to propagate the transgenic animal genetics into commercial dairy herds. Rapid dissemination of the genetics of the parental animals by nuclear transfer could result in the generation of mini-herds in two to three years. However, the existing inefficiencies in nuclear transfer make this a difficult undertaking. It is noteworthy that the genetic merit of the 'cloned' animals will be fixed, while continuous genetic improvements will be introduced in commercial herds by using artificial insemination breeding programs (Karatzas, 2003).

In an alternative scenario of herd expansion, semen homozygous for the transgene may be available in four to five years. Extensive breeding programs will be critical in studying the interaction and co-adaptation of the transgene(s), with the background polygenes controlling milk production and composition. Controlling inbreeding and confirming the absence of deleterious traits so that the immediate genetic variability introduced by transgenesis is transformed into the greatest possible genetic progress is equally critical (Karatzas, 2003). Another alternative strategy for transgenesis is based on the use of sperms as vectors in the integration of the transgenes. Initially described in mice (Lavitrano et al. 1989). Results showed that this procedure might be efficient in sheep (Niemann, 1998). In addition, a successful expression of a gene related to genetic modification of pigs for a gene related to xenotranplantation was obtained using this technique.

Eighty percent of the pigs were transformed and 54% expressed the transgene consistently (Lavitrano et al. 2002). A very efficient modification of this technique that uses the co-injection of sperms and DNA, has been described in the mouse and given a high rate of transgenesis (20%), therefore, their application to domestic species seems promising. Intracytoplasmic sperm injection (ICSI) has been used recently for the stable incorporation and phenotypic expression of large yeast artificial chromosome (YAC) constructs of submegabase and megabase magnitude. This technique allowed for more than 35% of transgenesis (Moreira et al. 2004). Another option for transgenesis is the use of insertional mutagenesis using natural transposons. A transposon system called "Sleeping Beauty", and active in a wide range of vertebrate cells, was used to transform mouse embryos with mRNA expressing the SB10 transposes enzyme (Dupuy et al. 2002). Kuroiwa et al. (2004) targeted sequentially a system for primary fibroblasts cells that were used to knock out both alleles of a silent gene, the bovine gene encoding immunoglobulin-μ (IGHM), and the active gene encoding the bovine prion protein (*PRNP*) and produced both heterozygous and homozygous knockout calves. The procedure integrates homologous recombination to replace genes in cell culture, and rejuvenation of cell lines by production of cloned fetuses. A method for selective elimination of selection marker genes was also developed. This method allow for the production of double homozygous transgenic embryos in 21.5 months. In contrast, for cattle, the production of double homozygous from heterozygous founders would require approximately 5 years and generation for double homozygous from heterozygous founders is impractical. This method can be used to breed many types of cattle with improved disease resistance and values for increased productivity. A recent alternative consists on the transformation of somatic tissues of developed animals, using techniques similar to those used in gene therapy (Kinghorn, 2003).

Discussion

Detecting genes related to disease and their expression in humans from studies on the genome, could lead to the development of therapies and the development of drugs for specific individuals, and enhanced early diagnosis of individuals with high-risk genotypes, allowing for preventive or remedial actions, even gene therapy. In animals, this knowledge could lead, in addition, to select against defective genes.

In livestock, knowledge of effects of specific genes and gene combinations on important traits could lead to their enhanced control to create new, more useful populations. The use of specific gene information is not a panacea, but could help to increase rates of genetic improvement, and open opportunities for using additive and non-additive genetic effects of domestic species, provided wise improvement goals are used and this new technology is optimally used together with the so called 'traditional' or 'conventional' methods based on phenotypic and genealogical information.

These methods will help to increase our knowledge about the genetic architecture of complex quantitative traits in domestic animal populations and to estimate the distribution of the genetic variation across and within breeds and population. It will also aid in ascertaining the genetic merit of local, less known populations (Hill, 2000). Studies for using genetic diversity in structured populations using DNA markers (Hartl and Clark, 1997) are very useful in order to set priorities for conservation of distant or unique populations as reservoirs of potentially unique genes, because their contribution to biodiversity would be greater (Oldenbroek, 1999). Currently, however, the main practical application of DNA markers is for parenting determination and to trace products such as meat. Despite its relatively low success rates and associated high costs, transgenic technology have a number of important potential applications in animal improvement such as increasing productivity, product quality and creating novel products. A major limitation to use transgenesis in the improvement of productive characters is the limited knowledge available on the identity and regulation of the genes that control these characters. The advance in the elaboration of genetic maps and fine positional cloning studies in the main species of interest will allow having a larger number of candidate genes susceptible of being manipulated. However, the road from genotype to phenotype is proving to be much more complex than previously thought for disease and production traits affected by many genes (True et al. 2004).

One promising applications of transgenesis is the synthesis of biomedical products of high commercial interest. Transgenic bioreactors and the use of exogenous or artificial genes interfering with particular cell mechanisms or with pathogens but not, or only marginally, with the physiology of the animals are potential applications. A greater knowledge on the mechanisms that determine the integration of the transgenes and genic regulation will allow a more precise control of the expression of the transgenes and it will probably facilitate a larger number of applications in the domestic species, including modifications beyond normal limits, such as to increase the number of copies of the gene and their expression. These transformations could be regarded as a form of mutation (Hill, 2000). The expressions of complex traits are the result of several mechanisms involving both regulatory and structural portions of the genome. Advances in molecular genetics, genomics, proteomics and transcriptomics might perhaps help to shorten the gap between the more 'holistic' approaches of quantitative genetics with the more 'reductionistic' approach of molecular genetics. The release of genome sequence information in cattle (Sonstegard and Van Tassell, 2004) and pig (Wernersson et al. 2005), may allow for a more efficient use of MAS and also to address some consumers concerns regarding product quality and safety.

Use of genetic engineering for animal and plant improvement is in its infancy, therefore many questions regarding efficiency, safety and societal benefits in particular situations remain. Problems arising transgenic plants, including their lower-than expected productivity, are reviewed thoroughly by McAfee (2003). Simplistic and overoptimistic views of biotechnology should be replaced by serious and scientifically based assessments of these new technologies by potential users on a case-by-case basis. We need to emphasize that in most cases, the use of MAS is not a revolution but just an evolution with regard to the traditional methods, because we are looking to improve more efficiently traits that already are actually or potentially improved in an efficient way using, for instance, mixed model (BLUP) based technologies for selection. Efficiency issues are very important. In order to increasing the efficiency of MAS, we need previously to:

1. Define with greater precision the selection goal and selection criteria (Monin, 2003).

2. Optimize the use of BLUP and other 'classical' breeding methodology.

The use of transgenic animals models for the study of gene regulation and expression has become commonplace in the biological sciences. Contrary to the early prospects related to commercial exploitation in agriculture, there are some challenges regarding their use that still lay ahead. The risks at hand can be defined not only by scientific evidence but also in relation to public concern (whether perceived or real) that exists in some people (Larrère, 2003). Therefore, the central questions will revolve around the proper safeguards to employ and the development of a coherent and unified regulation of the technology.

Cloning is another technique that raises concerns both from the ethical and practical point of view. Whether it is acceptable to clone humans is a very difficult issue. In animals, besides the very low success rates, some abnormalities should suggest that more information is required on the consequences of such practices in humans but also in animals, before its routine use. Advantages for animal breeding programs derived from cloning with no use of transgenesis are like to be small (Van Vleck, 1999).

These two examples illustrate that in spite most of the problems are technical in nature, implications of the use of this knowledge will be important for the society as a whole (Olsson and Sandoe, 2004).

A reasonable degree of regulation, open information on the issues of genetic engineering technologies from the academic world and an involvement of the whole society in the developments of the laws concerning these issues, seems to be the best way to circumvent an exaggerated or negative reactions to some of these knowledge, and to avoid or reduce unethical or abusive use of these techniques (Fukuyama and Stock, 2002). A specific set of conclusions regarding safety of food from genetically modified animals is available from a FAO/WHO expert consultation panel (FAO/WHO, 2003).

Concluding Remarks

Most of the important potential technical advances offered by genetic engineering technology in animal breeding are still ahead. Their use has both advantages and problems. Advantages are related to a more complete control over the animal genome. Problems are related to technical complexity, high costs, in some cases, public acceptance and ethical dilemmas.

It is not likely that this technology, will replace 'conventional' methods for genetic improvement. Instead, they probably will begin to be gradually incorporated into current genetic improvement programs that use efficiently classical improvement methods to achieve particular objectives.

Bioland Standards

"No act contrary to nature remains without consequences. No natural principle can be breached without its being punished, no natural order of things be dispensed with without danger to ourselves. The integration of humans in the order of creation is a vital prerequisite for their lives." —Dr. H. P. Rusch

Dr. Hans Müller and Dr. Hans Peter Rusch, in their work on the care of the soil and the maintenance of its long-term fertility, established the organic biological method of farming. This is based on the exact observation of biological connections of the effects between soil – plants – animals and humans with the aim of achieving optimum care of biological regulation systems in the agricultural field. Agricultural products are generated within as closed a business operating cycle as possible in the sense of a true original production. The mutual tasks of organic biological cultivation consist of:

- caring for the natural basics of the life of the soil, water and air
- producing foodstuffs of a high health value
- carrying out active nature protection and the preservation of species
- avoiding to damage the environment
- keeping animals according to the needs of its species
- making a contribution towards solving the world-wide energy and raw materials problems
- creating the basis for the maintenance and development of independent farming structures.

For decades farmers have been working according to the knowledge gained by Dr. Müller and Dr. Rusch and have mutually developed this further in their practical work. It has thus been possible for them in their fields of work to counteract the negative effects of the agricultural and social politics, to operate an environmentally friendly form of agriculture and, in co-operation with processors and consumers, to put a stop to the destruction of the existence of farmers. These farmers, gardeners, wine-growers and beekeepers in the Federal Republic of Germany combined to establish the BIOLAND e.V. Verband für organisch-biologischen Landbaumethoden (further in the text called BIOLAND) and have compiled the following standards.

The standards explain in detail the application of the organic biological methods of farming, the conversion to this method of operation and enable control of the cultivation defined according to the standards to be executed. It remains the mutual task of the people connected with BIOLAND to continue to work towards the aim of maintaining our natural basics of life and to improve the standards to keep them in line with the latest knowledge available.

EU-Regulation "Organic Agriculture"

During the drafting process of these standards the "EU Regulation on Organic Production of agricultural Products and Indications Referring thereto on Agricultural Products and Foodstuffs (EEC) Nr. 2092/91" and its amendments have been observed. BIOLAND member farms and contract partners are always obligated to adhere to the provisions of this EU-regulation in its currently amended form.

Genetic Engineering

Exclusion of Genetic Engineering

Genetically modified organisms (GMOs) and products derived therefrom are not compatible with the organic production method. Products, produced according to BIOLAND-standards, have to be produced without the use of genetically modified organisms (GMOs) and/or GMO derivatives.

Definition of Terms

'Genetically modified organism' (GMO) shall mean any organism as defined in Article 2 of Council Directive 90/220/EEC of 23 April 1990 on the deliberate release into the environment of genetically modified organisms 'GMO derivative' shall mean any substance which is either produced from or produced by GMOs, but does not contain them. 'Use of GMOs and GMO derivatives' shall mean use thereof as food stuffs, food ingredients (including additives and flavourings), processing aids (including extraction solvents), feeding stuffs, compound feeding stuffs, feed materials, feed additives, processing aids for feeding stuffs, certain products used in animal nutrition (under Directive 82/471/EEC), plant protection products, veterinary medicinal products, fertilisers, soil conditioners, seeds, vegetative reproductive material and livestock.

Crop production

Soil Fertility

The care of the soil and, correspondingly, the maintenance and the improvement of soil fertility constitutes a special point of emphasis in organic biological farming. A healthy, invigorated soil is the best prerequisite for healthy plants, healthy animals and healthy people. All measures of plant growing should form the basis for the improvement and care of a diverse and active soil life. Only the vitality of the soil itself will ensure long-lasting fertility.

Location

Ecological Design

In order to promote the health and the resistance of plants, the location must be designed in accordance with ecological points of view. For example, by planting and maintaining hedges, creating nesting possibilities and ensuring provision of shelter for insects, beneficial animals are to be encouraged and the self-regulation within the ecological system improved.

Selection of Location

In the choice of the location, the load created by pollutants from the environment and from previous usage of the soil are to be taken into consideration. If there is the danger of such a load being present, food stuffs and soil must be examined for residue.

Areas which have been affected by loads can only be used for organic biological farming when the loads involved have been reduced to a level which is justifiable for health by the adoption of suitable measures (e.g. protective planting). The BIOLAND Association can prohibit the use of the Association's name/trade mark BIOLAND on products which have been produced from land, partial land or border land contaminated by such loads.

Crop Rotation

Crop rotation is to be planned in such a variable and balanced manner that this fulfils the following functions:
*   the maintenance of soil fertility
*   the production of healthy plants
*   the suppression of weed in fields
*   the nutrition of animals using the business's own fodder

- the achievement of economically feasible yields without the use of chemical fertilisers and chemical products for plant protection.

In order to fulfil these functions, crop rotation must contain leguminous plants as main or intermediate crops or as mixed cultures.

Soil Preparation

The objective of soil preparation is the creation of optimum growth conditions for the crops. The compatibility with the soil life is to be taken into consideration in all measures adopted in soil preparation. Soil preparation must be carried out in such a manner that the natural soil structure is not excessively disturbed and that a loss of nutritional content and unnecessary expenditure of energy are avoided.

Fertilisation and Humus Management

Basic Principles

The objective of fertilisation is to achieve harmonic nutrition of the plants by means of a soil full of life. Organic material from the business itself forms the basis of fertilisation. It is mainly added to the soil by means of spread composting. Manure from the business itself must be prepared and spread in such way that the life in the soil is supported and the humus content is maintained or increased.

Permissible External Fertilisers

In order to complement the fertiliser produced in the business itself and to compensate any losses in nutrition caused by the operational cycle, fertilisers from external farms and organic and mineral fertilisers may be used in as far as these are listed in 10.1. Farm fertilisers from conventional sources must be subjected to careful composting.

They may only be used when they are considered harmless in regard to their pollution content. If necessary a quality examination can be requested. Trace elements may only be used when the deficiency determined cannot be removed by any other means.

Non-permissible Fertilisers

The use of farmyard slurry and urine and poultry manure from conventional animal farming is forbidden. In addition, the use of chemical synthetic nitrogenous fertilisers, easily soluble phosphates and other fertilisers not listed in 10.1 is prohibited.

Quantity Limitation

The total volume of organic fertiliser, based on the nitrogen content, may not exceed the amount which corresponds to an animal livestock count of 1.4 manure units (= DE) per ha. A maximum of 0.5 DE of this may be organic fertiliser from external sources.(DE = maximum animal stock density according 1.4 DE). The conditions specifie apply to gardening and perennial crops. In measuring the fertilising, the reserves available in the soil must be taken into consideration.

Production of Quality and Environmental Compatibility

Fertilising is to be designed in conformity with the location and the crops involved in such a way that the quality of the products (physiological nutritional value, taste, imperishability) may not be detrimentally affected in particular by the amount of nitrogenous fertiliser. In regard to the type, the amount and the time of applying fertiliser, care must be taken to avoid placing loads on the soil and the water (e.g. through heavy metals and nitrates).

Sewage Sludge and Compost

The use of sewage sludge and refuse compost is prohibited. Organic compost from separated collection and peat substitutes (e.g. bark products) may only be used after prior analysis of their pollutant content and following agreement with the BIOLAND Association.

Seeds, Seedlings and Plant Materials

For growing, those species and varieties of plants should be used which are best suited for the conditions prevailing at the location, they should not easily be subject to disease and be of a high physiological nutritional quality. In farming, varieties typical for the area should be used in preference to hybrid varieties.

Organically Produced Seeds and Plant Materials

When certified seeds and plant materials of suitable varieties are available from organic propagation, then these must be used. Any other sources require the express exceptional approval by the BIOLAND Association. From August 1st, 2006 on it is the objective to use seeds and plant materials from organic propagation exclusively.

Treatment of Seeds

Seeds and plant materials may not be treated after the harvest with chemical synthetic pesticides (e.g. disinfectants). Care is to be taken when using conditioned seeds (pelleted seeds, seed plates, etc.) to ensure that the materials used are harmless in the sense of these standards.

Seedlings

The seedlings used in the business must be grown by the business itself or be purchased from other farms of the BIOLAND Association, if here not available in accordance with the requirements of BIOLAND from other organically managed farms Substrates for cultivation may only contain a maximum of 80 vol. % of peat. Peat substitutes must low in pollution and ecologically compatible.

Young Plants for Perennial Crops

Young plants used in the business must be purchased from nurseries or producers of propagating materials of the BIOLAND Association, if here not available in accordance with the requirements of BIOLAND from other organically managed farms, if the required varieties and suitable qualities are available. Other sources require express approval by the BIOLAND association.

Plant Protection

The objective of organic-biological farming is to produce plants under such conditions that their infestation with parasites and disease achieves a point where this is of no or only minor economic significance. Appropriate measures for the achievement of this are balanced crop rotation, selection of suitable varieties, soil preparation in accordance with the location and the time of year, fertilising in appropriate amounts and qualities, fertilising by growing, etc.

In addition, the spread of beneficial animals should be promoted by suitable means and measures such as hedges, nesting places, wet biotopes, etc. If equipment is used that is also used to disperse non-permissible plant protection agents according to the Bioland standards, prior to the use it has to be drained and cleaned thoroughly.

Permissible Measures

Special preventive measures should only be carried out using the agents which are listed in They are only to be used when all other measures for activating the defensive powers of the soil and the plants themselves and the design of the location have been exhausted. The legal regulations regarding their use are to be observed in using plant treatment agents.

Prohibitions

The use of synthetic pesticides and growth regulators is forbidden.

Weed Regulation

The regulation of weeds is effected by preventive measures (e.g. crop rotation, soil preparation, variety selection), mechanical measures (e.g. harrowing, raking, hoeing) and thermal measures (e.g. burning off).

Prohibition of Herbicides

The use of herbicides is forbidden.

Air, Soil and Water Protection

Water resources are not to be used excessively. It is not permissible to burn used plastic (e.g. foils and fleeces) in the fields.

Wild Collection

The collection of edible plants or parts thereof, growing naturally in natural areas or forests, where the only humen interference consists of the hearvest (collecting) of the products, is considered as wild collection, if the following conditions are observed:

- The collecting area must be clearly defined. It must be identified by the way of land register maps (if necessary drawing of plans).
- The collection in areas out of the area under the care of the BIOLAND Association is only allowed with prior approval.
- The collecting area shall not be under the direct influence of any sources of pollution.
- The areas have received no treatments with products other than those allowed by this standards (Annex 10.1 and 10.2) for a periode of three years prior to the collection. This must be documented by appropriate means.
- The collection shall not affect the stability of the natural habitat or the maintenance of the species in the collection area.

Those products may be labelled with the trade mark/association name of BIOLAND with the addition "... from wild collection" (processed products in the list of ingrediences).

Animal Husbandry

Keeping animals is a sensible link in the operational cycle. Keeping animals in accordance with the needs of the species and their considerate care by humans form the prerequisite for the health, the performance and the well-being of the animals. With the help of the animals, the feed produced in the operation is used in the production of foodstuffs of a high quality for human consumption. Animal keeping is to be designed in such a way that it can be assured that the production, storage and spreading of the manures produced in the operation of business through keeping animals is with as little loss as possible. This serves to maintain and improve the fertility of the soil in the business.

Requirements in the Keeping of Animals

Keeping animals in accordance with the needs of the species must be the objective of every business. This means that the behaviour peculiar to the species in question, such as behaviour in movement, rest, ingestion of feed, social contact, comfort and reproduction is made possible as far as feasible. To promote robustness and vitality, the animals should often be allowed to face the weather and climatic conditions of the location.

Keeping animals in a manner peculiar to the species involves providing sufficient space for movement and rest throughout the year, natural light, shade, protection against wind, fresh air and fresh water. It is mandatory for all animals to have access to open air and/or grazing (for existing livestock buildings transitional periods until 2010 are possible for poultry, pigs and young cattle with permission of the BIOLAND Association, compare 9.4). Injuries and illnesses resulting from their being kept must be avoided. Herd animals may not be kept individually. Keeping animals individually is permissible only in case of male animals for breeding purposes, in case of sickness, versus the end of pregnancy and in small stocks.

Space Requirements

Requirements for in-and outdoor areas of the livestock housing system are listed for each type of animal in attachment no. 10.6 (for existing livestock buildings transitional periods until 2010 are possible with permission of the BIOLAND Association, compare 9.4). For the keeping of fallow-deer and red deer the regulations. Housing systems for mammals with no clear separation between in-and outdoor area have to fulfil space requirements in sum. In case of housing systems for ruminants and horses with free range barn and access to pasture in summer, the space requirements for outdoor area according to attachment 10.6 can be skipped. In this case permanently accessible, durable and non-roofed barn area can be taken into calculation of the total barn area.

In regions with suitable climatic conditions, allowing animals to be kept outdoors all year round, housing is not stipulated.

Movement and Rest Area

Barns with fully perforated floor area (fully slatted floors, flat decks, cages) are not permissible. The width of slots and holes in case of perforated floors have to be adapted to animal size. Slatted floors must be in excellent technical condition. Surface slats are to be preferred. The majority of accessible movement and rest area for each mammal category must be a solid floor area (no slat floors). The tread surface must be non-slip and of tread proof nature.

A soft, dry and clean place in which to rest is to be ensured at all times for ruminants, pigs, horses and rabbits by means of strewing (as a rule, straw). Straw for bedding purposes should, as far as available, be from the farm itself or from other organic farms. Conventional straw for strewing should be grown on lands with a minor degree of farming intensity.

Assessment of Housing Systems

As an orientation aid in evaluating the adequacy of the compliance in the keeping of animals, the index of adequacy in animal keeping (TGI = Tiergerechtigkeitsindex, prepared by the society for ecological animal keeping (GÖD = Gesellschaft für ökologische Tierhaltung)) can be applied. New buildings and alterations to existing buildings for keeping animals should be in accordance with the latest status of knowledge in regard to keeping animals in compliance with the needs of their species. The planning of such buildings should therefore be co-ordinated with the BIOLAND Association.

Access and Care of Open Air Run Areas

Access to open air run and pasture has to be afforded always when physiological, climatic and soil conditions allow for this. The stock density of animals on outdoor areas may not result in the soil being trampled down – with the exception of feeding and drinking areas. Over-grazing has to be avoided.

Construction and Maintenance of Livestock Buildings

In construction and maintenance of livestock buildings ecological issues have to be considered. Substances hazardous to health and environment in building materials and its treatments have to be avoided, if possible.

Native building materials have to be preferred.

The use of non-regenerative energy resources in the construction and maintenance of stables has to be reduced, if possible.

Keeping Cattle

Dairy Cattle and Suckler Cow Keeping: Cows are afforded access to meadows or are allowed out into the open air at least during the six summer months.

Cows should be afforded the possibility to calve in separated calving booths or on the pasture.

Non-Penned Cowsheds

Efforts should be made to have non-penned cowsheds which allow the cattle freedom of movement.

Dead ends and bottle necks in the non-penned cowsheds are to be avoided.

If outdoor grazing in summer is not possible, then access to open air has to be afforded all year round (for existing livestock buildings a transitional period until 2010 is possible with the permission of the BIOLAND Association, compare 9.4).

In winter the possibility of regular movement in the open air should also be afforded. There must be a place in the non-penned sheds for each animal to sleep and eat. A slight reduction in number of eating places is possible with the permission of the BIOLAND Association in case of permanent availability of fodder (storage feeding system).

Boxes in which the animals can rest must enable the animal to lie down and rise up in a manner in compliance with the species. Keeping animals tethered in combination with summer pasture and/ or regular access to open air is possible with the permission of the BIOLAND Association during a transitional period until

2010 in existing livestock housing (compare 9.4). Tethering system exceeding 2010 is possible for small animal stocks if cows are afforded access to open air, at least twice a week. Tethering of single animals for security or animal protection reason is possible with the permission of the BIOLAND Association as long as it is limited in time. If the animals are kept tethered, the width, the length and the technology used in tethering and thedesign of the edges of the trough must allow the animal to stand up, lie down or eat in a manner suitable for the species and must allow the animal sufficient body care.

The cows must be able to fully stand and rest on the level, secure surface, which has to be strewn sufficiently. Rigid neck frames and tightly drawn chains or nylon belts are not permissible. Electric cattle trainers are forbidden.

Cattle for Breeding and Beef Cattle

All cattle for breeding purposes and for beef should have the possibility of free movement throughout the whole year. Efforts should be made for grazing throughout the vegetation period. If grazing in pastures is not possible, cattle for breeding and beef can be kept throughout the year in non-penned cowsheds (for existing buildings a transitional period until 2010 is possible with the permission of the BIOLAND Association, compare 9.4). Keeping beef cattle without access to open air is permissible only during the end of the fattening period to a maximum of 1/5 of the life time, but in any case no longer than 3 months. If grazing possibilities are afforded during the vegetation period, tethering is permissible for breeding and beef cattle of an age of over 1 year.

Calves

The calves should be able to stay with the mother for at least 1 day following the birth. From their second week of life, when the number kept is correspondingly large, the calves must be kept in groups. In case of their being kept in cowsheds (huts/igloos) with the respective generous possibility to move about and to have social and eye contact, they can be kept in groups from the 6th week of life on. Tethering of calves and young cattle under the age of 1 year is not permissible.

Keeping Pigs

Pigs must be allowed access to open air runs (for existing livestock buildings transitional periods until 2010 are possible with the permission of the BIOLAND Association). With the exception of the late pregnancy period and the suckling period of sows, pigs have to be kept in groups. Tethering of sows is not permissible. Fixing should only be undertaken with problematic animals during and after farrowing. A wallowing area should be available. Housing pigs without access to open air is only permissible during the final fattening period for a maximum of 1/5 of life time, but in any case no longer than 3 months. During the 6 summer months period breeding pigs, wherever possible, are to be afforded access to a pasture. The pasture should have shady areas and an area for wallowing.

Sheep and Goats

The stables must be designed in a non-penned manner. During vegetation period efforts must be made to keep them on pastures. If no pastures are afforded, sheep and goats have to be kept in non-penned stables with access to open air. During final fattening period the keeping of fattening lambs and goats is only permissible to a maximum of 1/5 of the life time, but in any case no longer than 3 months.

Barn

Keeping laying hens in barns is in the form of floor or volary systems with access to open air run. The single barns with a maximum of 3000 laying hens have to be separated in such a way, that infections and/or a contamination with parasites will be reduced, and to ensure a sustainable management of the greened area for movement (for existing barns transitional periods until 2010 are possible with the permission of the BIOLAND Association, compare 9.4). Per each m^2 of movement area in the barn accessible to the animals, 6 animals can be kept. Movement area that accounts for the calculation of the animal stock density has to fulfil the following requirements:

- minimum width at least 30 cm
- maximum slope 40 %
- in case of gritted floor minimum wire strength to be kept at 2 mm
- free height between floor levels or perch rods at least 45 cm
- durable floor area has to be covered with suitable strewing material in sufficient thickness
- laying nests, its landing grids and higher perch rods are no moving areas and can thus not be accounted for in animal density calculations.

The exterior climate area can be accounted into accessible barn area, if:

- it is accessible through all barn openings for all animals during their total activity time (light phase, natural and artificial light)
- if it is roofed and equipped with automatic opening system, lighting, fencing and wind protection possibility (only in very cool temperatures and strong winds the number of barn openings can be reduced)
- if the whole exterior area is strewn with sand or similar for all animals
- if it has a height of at least 2 m
- if it is located on the same level as the barn; a level difference between barn and exterior area is limited to a maximum of 50 cm (in case of higher level differences, a sufficient circulation of animals can be reached by the construction of balconies and climbing and descending supports).

In relation to the barn floor area the maximum animal density in volary systems is 12 animals/m². In barns with an integrated exterior climate area, a maximum of 8 laying hens/m² accessible area in exterior barn areas can be kept at night, in volary systems a maximum 15 animals/m² floor area.

The barn has to be designed in such a way that animals will have the least possible contact with excrement. The different floor levels accessible to laying hens have to be arranged in such a way that excrement do not fall on the level beneath it. At least 1/3 of the moving area in the barn for all animals must be available as strewn area for the purpose of scratching. In barns with integrated exterior climate area this third refers to the interior area of the barn. The strewing material has to be at least 5 cm deep and must be kept dry, loose and clean. The barn has to be lighted sufficiently with day light. The window area has to be equivalent to at least 5 % of the barn floor area. Natural day light may be extended to a maximum of 16 hours by artificial light. The offered space for feeding, feeding dishes and the strewn area for the application of feeding seeds have to be designed in such a way, that all animals can forage together.

The animals should be able to take water from an open water surface. The available drinking water has to be always fresh and clean. At least 18 cm of perch rod have to be provided. In barns with excrement pits at least 1/3 of the perch rods have to be elevated at least 45 cm. The profile of the perch rods must have at least 30 x 30 mm, the upper edges of the rods have to be rounded. For the total perch rod length only such perch rods are accounted for, that have at least 30 cm horizontal axis distance from each other and at least 20 cm distance from the walls. For the laying of the eggs the animals must have available sufficient strewn laying nests or rolling nests with smooth rubber nops or similar material. For 80 laying hens 1 m² family nest has to be available, a single nest is sufficient for a maximum of 5 hens.

The animals must have permanent access to a dust bath, if possible in a winter garden. The barn openings to the exterior climate area and the access to the outdoors are to be measured in such a way that the animals can circulate without problems and unrestricted. The barn openings have to add up to a combined length of 4 m per 100 m² of the floor area of the building available to the hens (for existing buildings a transitional period until 2010 is possible with the permission of the BIOLAND Association, compare 9.4). The minimum measures of the openings are 50 cm width and 45 cm free height. Between charging the barn has to be cleaned and disinfected. Only the substances listed in attachment 10.7 are permissible.

Exterior Climate Area (Winter Garden)

For an animal stock density of more than 4 hens per m² in the barn at least 1 m² of durable exterior climate area per 12 hens is mandatory. Excepted from this are stocks of less than 200 hens and mobile barns. The exterior climate area can be accounted for in the calculation of the animal density provided they fulfil the conditions.

Green Open Air Run

A green open air run is mandatory (for existing buildings transitional periods until 2010 are possible with the permission of the BIOLAND Association, compare 9.4). To each animal at least 4 m² of green roaming area in a perimeter of 150 m has to be available. Measures have to be taken that a nutrient intake of 170 kg N per ha roaming area and year must not be exceeded. Strongly used areas close to the barn are to be strewn with bark shred or similar and designed in such a way that the strewing, soil material respectively, enriched with nutrients, can be replaced periodically, latest before next recharging of the barn. Plants must grow on the majority of the outside roaming area. Frequent and sufficient resting periods have to be scheduled for the regeneration of vegetation.

Access to the green roaming area has to be provided during the whole vegetation period from 12.00 h noon on. In extreme weather conditions (snow, permanent rain, thunderstorms) access to green roaming area can be restricted in time or totally. The green roaming area has to offer protection from enemies and shade to the animals, so that they will use the roaming area in an equally distributed manner. Thickets will be planted for a natural structure of the whole roaming area. Shading or wind protection nets will provide artificial possibilities of shelter.

Young Hens

The regulations on laying hens as described above apply for young hens respectively as far as in the following paragrafs no other regulations to be met. Additionally the following has to be observed:

Principle

During growth the young animals should learn the natural behaviour which they can conduct in the laying barn, that way avoiding behavioural disorder. During growth robustness should be Bioland Standards April 26th, 2005 page 12/12 developed and a natural immunisation should be achieved. The housing system in the growing barn should be equivalent to the barn of the laying hens.

Barn

In the prime weeks of life rings for chicken are permissible. From the 3rd up to the 12th week of life max. 15 kg life weight and not more than 16 animals per m² moving area are allowed to be kept. From the 12th week of life per m² of accessible moving area a maximum of 10 animals can be kept in the barn.

In barns with multiple levels a maximum to 24 animals per m² barn floor area can be kept from the 12th week of life on. In barns with an integrated exterior climate area in the 12th week of life, at night times, a maximum of 13 young hens per m² accessible moving area can be kept in the heated area, provided the exterior climate area is permanently accessible during the light phase. At the earliest from the 6th life week on in addition the exterior climate area (wintergarden) may be taken into account. At least half of the moving area of the barn must be provided as area for scratching purposes. The strewing material has to be at least 5 cm deep and to be kept loose, dry and clean. Daylight with natural intensity is obligatory. If suitable equipment is installed, the application of a lighting program can limit light exposure and period.

Clean and fresh drinking water is always supplied to all animals. The equipement for feeding shall be constructed in a way that all annimals are able to eat at the same time. From the 1st week of life on the animals must be provided with raising possibilities, from the 8th life week on 8 cm per animal, from the 12th week on 12 cm of perching rod per animal is required, whereas 1/3 are to be designed as elevated perching rods. From the 1st week of life on the animals must have available a dust bath and strewing material with sand and coarse-graind limestone material as well as opportunities for shelter and to cover.

Exterior Climate Area and Outdoor Run

From the 10th week of life latest the animals must have access during their activity period to a durable, roofed exterior climate area in the size of at least one quarter of the accessible barn area, depending on the extent of feathering and the climate. The size of the barn openings is at least 2 m per 1000 young hens.

Stock sizes of less than 200 young hens or mobile stables are excluded, if a green open air run of at least 2,5 m² per animal is available.

It shall be possible to divide the green outdoor run into portions, and it must contain equipment for shelter

Poultry for Fattening

The regulations on the keeping of laying hens apply for the keeping of poultry for fattening respectively. Additionally the following applies:

Barn

Locally the single barn buildings with a maximum of 4800 fattening chicken, 5200 guinea-fowls, 4000 female ducks, 3200 male ducks or 2500 geese and turkey hens have to be separated in such a way that infections and/or contamination with parasites are reduced, and a sustainable management of the green roaming area is achieved (for existing barns transitional period until 2010 is possible with the permission of the BIOLAND Association). The total used area of all fattening poultry barns of one single business may not exceed 1600 m² (for existing barns transitional period until 2010 is possible with the permission of an inspection authority, compare 9.4). In the barn per m² of accessible roaming area a maximum of 21 kg live weight and not more than 10 animals can be kept, whichever case applies first. Roaming area is defined as the floor area of the barn available to the animals. In mobile barns a maximum of 30 kg live weight and not more than 16 animals can be kept per m². For guinea-fowls a minimum of 20 cm perching rod per animal is mandatory. For fattening chicken and turkey hens perching rods in respect to their size and age are to be offered.

Exterior Climate Area

For fattening chicken and turkey hens an exterior climate area or a durable roaming area is mandatory in addition to the interior barn area. The size has to be at least one third of the minimum barn area. Exempt from this are stock sizes of less than 100 animals and mobile barns. The exterior climate area can be taken into the calculation of the animal density.

Greened Roaming Area

A greened roaming area is mandatory (for existing barns for fattening chicken with respect to the afforded greened roaming area, for other fattening poultry with respect to the size of the greened roaming area respectively, transitional period until 2010 is possible with the permission of the BIOLAND Association, compare 9.4). To every animal the following minimum greened roaming area per animal has to be afforded:

- fattening chicken and guinea-fowls 4,0 m²
- ducks 4,5 m²
- turkey hens 10 m²
- geese 15 m²
- fattening poultry in mobile barns 2,5 m².

If the climatic conditions and the physiological conditions of the animals allow for this, access to a greened roaming area has always to be afforded, nevertheless if possible at least during one third of the lifetime. Restrictions may result from the physiology by the age and by the feathering of the animals and from the climate.

Water Surfaces

Water fowl has to be afforded access at any time to running streams, ponds or lakes (only if hygienic conditions and water protection acts permit it) or to a durable water surface that is replaced regularly by fresh water.

Keeping Horses

Whenever soil conditions allow, horses have to be afforded grazing or exterior roaming. For them being kept in barns it has to be in the form of boxes or non-penned stables with access to open air roaming area, if possible. As far as possible, the animals are to be kept in groups.

Keeping of Fallow-Deer and Red Deer

For fallow-deer and red deer keeping on pasture is mandatory all year round. The minimum preserve size for fallow-deer is 3 ha, for red deer 5 ha. In the pens there must be hiding places for calves. The pens must provide shelter against climate conditions, preferably by means of natural hedges and trees. Red deer pens must have a slough additionally. The minimum pack size of deer comprises of 5 animals (1 stag, 4 females). Per ha preserve area the stock density is 7 PED or 3.5 PER respectively. One production unit fallow-deer (PED) comprises of 1 adult animal, 1 calf, 1 one-year-old, one stag proportionately; one production unit red deer (PER) comprises equally.

Keeping of Rabbits

The following regulations apply for keeping systems of more than 3 animals for breeding purposes or more than 20 animals for fattening respectively.

General

Barn and roaming area allow for the kind of behaviour peculiar to the species. Keeping in groups is mandatory except during the nursing time of females. The maximum group size is limited to 40 animals for fattening purpose or 10 animals for breeding purposes respectively.

Barn

Daylight inside the barn is mandatory. The height of the barn must be at least 60 cm. The area of movement can spread on multiple levels. It ought to contain different surface features. Possibilities for retreat and resting areas must be available for all animals. Each nursing female needs its own nest to litter.

Roaming area

A durable roaming area is prescribed except during times of nursing and in the case of mobile barns. If a greened open air run is provided, areas for rotation and resting periods for the vegetation is mandatory for the run to remain mostly greened and to keep the load of parasites small. In the case of mobile barns there have to be greened open air runs.

Dealing with Animals

General, Dealing with animals must take into consideration the needs of the species and the feelings of the animals.

Measures in the Business

Care of hair, skin and hooves is to be carried out at regular intervals. As far as this is possible in the system used in keeping, no removal of horns should be carried out in the case of ruminants.

Not permissible are:

- removal of horns by means of cautery sticks
- cropping of tails in the case of cattle or pigs
- prophylactic shortening of pigs' teeth
- insertion of nose rings and nose clamps to-prevent pigs from grovelling
- disfigurement of poultry by shortening of the beaks, cropping of comps and wings.

Animals must not be subjected to further surgical interference systematically. Keeping a laying pause is possible for laying hens. During this resting period the free access to water and fodder may not be restricted. The day light may be limited to 5 hours per day. As far as possible in each flock at least 1 cock per 100 hens should be kept from the beginning of the rearing time.

Animal Density and Purchase of Additional Feedstuffs General

The animal density is oriented in the main on the basis of the provision of feed by the business itself. In case of additional purchase of feed, at least 50% of the feed for one animal species must come from the business itself or from other farms of the BIOLAND Association, if here not available in accordance with the requirements of BIOLAND from other organically managed farms, or, if here not available, according to the regulations for the purchase of permissible conventional feed.

Ready mixed feedstuff shall be purchased only from feedstuff companies certified by BIOLAND, mineral feedstuff only from companies approved by BIOLAND. In case of poultry, pigs and horses, 100 % of the fodder can be purchased in small stock sizes, if the stock size in the business is less than 1000 laying hens (or the respective figure of other poultry categories), less than 30 sows, less than 60 fattening pigs places or less than 10 horses and simultaneously, the total stock concentration (=animals per hectare) limit of the business is not exceeded. Before December 31st, 2010 latest, this provision shall be reviewed.

Feed from Land in Conversion

Feed produced on land in conversion (compare 9.2.3) may be used up to a maximum of 30 % in the ration, in relation to the annual average per animal category. If the feed produced from land in conversion is from the business itself, this percentage amounts to 60 %. In case of simultaneous conversion of the whole business, this percentage can be periodically higher (compare 9.2.4.2).

Green fodder, preserved green fodder and other forage from lands, which have been managed organically throughout the total vegetation period of the plantation, but the last conventional measure dates back less than 12 months, are to be regarded as fodder of non-organic origin. They can be used for all categories of animals without authorisation, taking into consideration the actual and respective maximum percentage of conventional feedstuff in the diet alowed per day and per year in the EU regulation 2092/91 (other categories of animals max. 10% in the annual ration and max. 25% in the daily diet referring to the dry weight content of the fodder with agricultural origin).

Quality of Purchased Feed

The purchase of fodder is subject to most stringent quality standards in order to minimise the load placed on the operational cycle of the business by pollutants. Imported feed from Third World countries, fodder of animal origin, extraction groats and spoiled feed may not be used.

Feed Additives

Permissible are in particular the mineral substances and additives necessary for a nutrition in compliance with the needs of the animals involved as well as harmless, natural feed additives helping to improve the business's own feed and the health of the animals. The use of feed with active substances or additives such as antibiotic, chemobiotic or hormonal performance boosters, coczidiostatica, histomonostatica, copper as performance booster, NPN-compounds, synthetic aminoacids and synthetic colorants is forbidden.

The used vitamins, trace elements and additives must be used from natural origin, as far as they are available in sufficient quantity and quality. If they are not available and need is determined, the individual feed and additives as listed in appendix 10.4.6 may be used in animal nutrition.

Feeding

In principle, the feeding of the animals is done with fodder of organic origin. Animal feeding is to be designed in such a way, that fodder produced in the business itself is to be used to achieve animal products of high value. Feeding in accordance with the needs of the species, in addition to the determination of rations according to the animal needs, also involves supplying feed as per animal's behavioural requirement. In case of non-availability and shortage of supply the temporary regulations for purchased feed from non-organic origin subject to authorisation.

Cattle Feeding

In cattle feeding basic fodder (straight fodder) from the business itself must be used. At least 60% of the dry matter in the daily ration must be roughage. In summer, the basic fodder has to be in its majority fresh green fodder. The rearing of calves is based on the business' own milk or milk from other farms of the BIOLAND Association or, if not available there, in accordance with the requirements of BIOLAND from other organically managed farms, for a period of at least 3 months. Feeding with hot air dried green fodder (cobs) should be waived wherever possible due to the high energy input required. This does not apply for milk producing businesses which, for reason of quality, cannot use silage.

Pig Feeding

Fattening pigs and breeding pigs are to be offered roughage

appropriate to their age. Piglets are to rear by natural milk for a period of at least 40 days.

Sheep and Goat Feeding

The basic principles of cattle feeding also apply for sheep and goats. The rearing is achieved with natural milk for a period of at least 45 days. In case of the rearing of lambs and kids in milk sheep or milk goat businesses as well as in problematic cases (outcast lambs, triplets, etc.) the use of organically produced cow milk or milk powder from organically produced milk is possible.

Poultry Feeding

At least 10 % of the daily feed ration for laying hens must be given as kernels (seeds) into the strewing material. Free consumption of mussel shells and grit or similar is to be ensured. The food ration must contain food components requiring little digestion (e.g. grass). For poultry in the fattening phase the feed must contain at least 65 % grain. Young hens must be able to take an appropriate kernels mixture out of the strewing material from the 7th week of life on latest. Water fowl (ducks, geese) should, from the 6th week of life on, be given part of their feed in moist form.

Horse Feeding

If in keeping horses in boarding stables fodder is used or treatment is effected by the horse owner, not being approved, it must be ensured that these are of no danger for the operational cycle of the business. The manure generated in this way is to be treated as organic manure from external sources.

Rabbit Feeding

The places of feeding must be accessible to all animals at all times. In the varied rations ingredients of firm consistency must be contained, e.g. beets, potatoes or tree branches.

Animal Health

The basis for the animal's health and fertility is suitable keeping, feeding and breeding. Preventive measures for maintaining the animal's health without the use of medication, for increasing the animal's own physical defence forces and to contribute towards avoiding illnesses are to be applied wherever possible. Hygienic measures as there are cleaning and Disinfection measures, keeping of resting times in nondurable open air runs and greened open air runs and measures of the pasture management are to be preferred to therapeutic treatment.

Treatment

If animals have to be treated, natural healing methods and homeopathic treatment are to be given priority. Conventional medication (ethical or on prescription) should be used exclusively to prevent unnecessary suffering on the part of the animal and to preserve life. These must be prescribed by the veterinary surgeon.

If one animal or a group of animals gets more than 3 treatments with chemical-synthetic allopathic medication per year or more than 1 treatment, if the productive life cycle is shorter than 1 year, then the animals or the products produced from them can no longer be labelled as organically produced or they have to restart the conversion period, provided the permission of the BIOLAND Association (compare 9.2.4). Exempted from this are vaccines, parasitic treatment and medication the use of which is regulated by official regulations. In case of the use of chemical-synthetic allopathic medication, the double period of waiting following their use as legally stated is to be observed.

If no legal waiting period is defined, at least 48 hours have to pass prior to the production of foodstuffs. The prophylactic use of conventional medication and hormones is forbidden. Exempted from this is medication the use of which is regulated by official regulations as well as vaccines. Within a veterinarian therapy hormones may be used in single animals. The use of synthetic substances that enhance growth or increase production or suppress natural growth are not permitted. Vaccines may only be used if diseases on that particular business are known to be problematic or it is to be expected that diseases may not be controlled by other management measures. Legally prescribed vaccines are permissible. The list of active substances and groups of pharmaceuticals the use of which is forbidden or limited must be observed when carrying out treatment (compare 10.5).

Stable/Barn Register

Detailed records are to be kept in a stable/barn register about all treatment of animals. This will include time of treatment, diagnosis, type and duration of treatment and waiting period for the medication used. The treated animals are to be labelled as such in a doubtless manner, in case of large animal each, in case of poultry and other small animals each or in groups.

Storage of Medication

Only medication the use of which is approved may be stored in the business. The medication is to be stored in a medicine cupboard safeguarded against access by unauthorised persons. A clear labelling of the medication has to be provided. Residual amounts of medication have to be disposed off appropriately.

Stable/Barn Hygiene

Environmentally compatible agents are to be used wherever possible in the cleaning and disinfection of milking machines and other barn equipment. The list of approved substances for cleaning and disinfection has to be observed (attachment 10.7).

Animal Breeding

Breeding must be planned in such a way, that the performance, the health, the vitality of the animals and the quality of the animal products are maintained and improved throughout different environmental conditions. Keeping domestic animals common to the region should, wherever possible, be promoted. In breeding animals and in the choice of the type and race of the animals, particular ecological requirements for location must be taken into consideration. In the case of breeding animals for milk production and for breeding purposes, especially the feature of longevity are to be observed. Types of animals and races not being suitable for the above described keeping systems (compare 4.2) must not be kept.

For fattening poultry the following applies:

- chicken 81 days
- Peking ducks 49 days
- female flying ducks 70 days
- male flying ducks 84 days
- "Mulard" ducks 92 days
- guinea fowl 94 days
- geese 140 days
- turkey hens 140 days.

Origin of Animals for Breeding-Purposes

The use of animals for breeding originating from embryo transfer should be avoided.

The breeding may not be carried out on the basis of animals permanently purchased from nonorganic origin.

Reproduction

Reproduction by means of natural copulation is preferable.

Artificial insemination can, however, be applied for the purpose of improvement of the breeding ability of the animals.

Embryo transfer is forbidden.

Additional Purchase of Animals Principles

The purchase of additional animals may only be from other farms of the Bioland Association or, if not available there, in accordance with the requirements of BIOLAND from other organically managed farms.

Only excepted from this are animals where it can be proven that their acquisition from such a business is not possible and where the BIOLAND Association has issued an exemption permit.

Possible Permits for Conventional Purchase of Animals Cattle

For the initial generation of a stock, calves for breeding purposes may be purchased after weaning, however latest up to an age of 6 months. Young cattle before the first calving and male cattle for breeding purposes may be purchased annually to an extent of 10 % of the adult cattle stock. For the purpose of massive expansion of the stock, the switch to another race, or the opening of a new business branch, or for breeds that are in danger to being lost to farming this percentage may be increased to 40 %. The use of the trademark/association's name "Bioland" is not permitted for beef from cattle that was born on conventional farms and/or raised with feedstuffs non-compliant with these standards.

Pigs

Piglets for breeding purposes and after weaning may be purchased up to a weight of 35 kg (until 31.07.2006). They should originate from a keeping system with strewing material. Young sows before having piglets for the first time and boars for breeding purposes may be purchased annually to an extent of 10 % of the stock of adult pigs. For the purpose of massive expansion of the stock, the switch to another race, the opening of a new business branch, or for breeds that are in danger to being lost to farming this percentage may be increased to 40 %.

Sheep and Goats

For an initial stock build-up female lambs and kids for breeding purposes may be purchased after weaning, however latest up to an age of 60 days.

Animals for breeding purposes may be purchased annually to an extent of 10 % of the stock of adult animals. For the purpose of massive expansion of the stock, the switch to another race, the opening of a new business branch, or for breeds that are in danger to being lost to farming this percentage may be increased to 40 %..

Poultry

Young hens may only be purchased, if they are raised on organically managed farms. These farms are obliged to follow the BIOLAND standards for rearing young hens. In stocks up to 100 laying hens young hens may be purchased up to the 18th week of life (only with permit of the control authority).

In the case of poultry for fattening purposes young poultry may be purchased that is less than three days old. Care should be taken to ensure that the choice of the race is suitable for the method of fattening with open air runs.

Rabbits

For the initial generation of a stock animals for breeding purposes may be purchased. Animals for breeding purposes may be purchased annually to an extent of 10 % of the stock of adult breeding animals. This percentage can be increased to 40 % in the case of massive expansion of the stock, switching to another race or opening a new line of production.

Marking of animals

All of the domestic animals kept on the business premises must be clearly identifiable. Thus all animals or groups of animals are to be marked and a register of animals must be kept.

Bee-Keeping

The general parts of the BIOLAND Standards are also applicable for bee-keeping in as far as there are no exceptions specified in the following. Beekeeping can be also carried out according to the BIOLAND Standards by businesses which do not cultivate any area under agricultural use.

Keeping of the Bees

Location of the Bee Colonies

Paragraph 3.2.2 of the Standards applies accordingly to the locations of the colonies. If the location of the hives is an agriculturally used field it must be managed organically. The location of the colonies has to be chosen in such a way that within a perimeter of 3 km an impediment worth mentioning of the bee products by agricultural or non-agricultural sources of pollution is not to be expected.

For pollen gaining it is not allowed to use crops of which the flowers were sprayed with pesticides. Also industrial areas or the vicinity to streets having huge volume of traffic (e.g. highways) should be avoided. If it is suspected that the load on the environment is too great, the bee products are to be examined.

If the suspicion proves to be founded, the location is to be abandoned. Only such numbers of bee colonies are to be placed in one location which allow adequate supplies of pollen, nectar and water for each colony. If canopies from cultivated plants are intended to be used, organically cultivated areas are to prefer as nectar collecting areas. The planned targeting of conventional intensive fruit cultures for nectar gathering or pollination is not permissible. The locations of the colonies are to be recorded in a movement plan throughout the year.

The movement plan must contain exact details in regard to period of time, location (fields, plots of land, or similar), canopy and number of colonies. Locations outside the area under the care of the BIOLAND Association are to be used only with permit. If locations of colonies are situated in areas, which are indicated by inspection authorities as to be unsuitable for organic apiculture, products from those areas must not be marketed with reference to organic production.

Hives

Hives must be constructed of wood, straw or clay. This does not apply to small parts, roofing, grid floors and feeding appliances. In manufacturing the hives pollutant free glues and paints (e.g. natural varnishes on a linseed oil or wood oil basis) are to be used. Varnishes containing pesticides or those manufactured in chemical synthetic processes are excluded. Treatment of the interior of the hive is prohibited unless this is done with beeswax, propolis or plant oils. Cleaning and disinfection is to be by means of heat (flame, hot water) or mechanical. In case of acute infections the use of NaOH-solution for the disinfection and cleaning of the hive with subsequent neutralisation by means of organic acids is permissible. The use of other chemicals is prohibited.

Wax and Honeycombs

The colonies are to be afforded the possibility of constructing natural honeycombs on several combs during the breeding season. Central walls, start strips may only be made of BIOLAND beeswax which has been produced in BIOLAND businesses from natural honeycombs or wax for decapping. Plastic central walls are forbidden. There may be no residue of chemotherapeutics which may indicate the non-permissible use of varroa or treatment against wax moth. Wax may not come into contact with bleaches or solvents or other additives. Only devices and containers of non-oxidising materials are to be used for the wax processing. For hive hygiene, only thermal processes, acetic acid or bacillus thuringiensis preparations are permissible.

Calming and Driving Away Bees

No chemical synthetic materials may be used to calm or drive away the bees.

Feeding Bees

The feeding of bees is permissible as long as this is necessary for the healthy development of the colonies. Within the scope of the possibilities of the business, bees should be fed using honey from the operation's own beekeeping. Feeding with sugar in any case requires an agreement by the inspection authorities and has to be limited to the winter hibernation period and for the creation of young colonies. Adulteration of the honey as a result of excessive winter feeding is to be avoided by removing this prior to the start of the gathering season. Gaps in the feed with nectar supply are to be filled only by BIOLAND honey. Feeding with pollen substitutes is not permissible. For the feeding only organically BIOLAND feedstuffs may be used, if not available, feedstuffs from other organic sources in accordance with the requirements of BIOLAND other organically produce managed farms.

Bee Health

The use of chemotherapeutic medication is forbidden. Only in combating the varroa mites is, in addition to the bio-technical and bio-physical methods, the use of permissible.

For bee colonies destined for the production of honey with their honeycombs, the use of these materials is only permitted in the timespan between the last honey harvest of these colonies and January 15th of the following year. Oxidation on metals where residues may be expected is to be avoided. All treatment measures used are to be recorded in a treatment journal.

Apicultural Methods

The curtailing of bee wings as well as other mutilations are forbidden. The larvae with drones may only be removed in order to fight a varroa infection.

Bee Breeding

The objective of the breeding is the keeping of varroa-tolerant bees in a manner suited to the ecological conditions. Natural breeding and reproduction processes are preferable. The swarm instinct is to be considered in this. Instrumental insemination may be applied in breeding businesses in individual cases if an exception has been approved by the BIOLAND Association.

Purchase of Additional Bees

The purchase of colonies, swarms or bee queens is only permitted from other businesses of the BIOLAND Association or, if not available there, in accordance with the requirements of BIOLAND from other organically managed farms. The catching of conventional swarms is permitted as long as its number does not exceed a limit of 10 % of the existing bee stock size in the business annually.

Honey

Only honey which has ripened in the hive may be extracted. Combs destined for the production of honey must not contain any eggs. The use of chemical repellents as well as the killing of bees during harvesting is forbidden. All harvesting measures have to be recorded in the colony register in combination with yield figures as exact as possible.

Processing

Warming of the honey should be carried out as gently as possible. It may not be heated to more than 40°C. The melitherm process is permissible. The honey should be filled wherever possible before it sets for the first time. Returnable sales units are prescribed.

To preserve the natural contents, the honey must be stored in dry, cool and darkened conditions.

To remove impurities such as wax parts, the honey may be passed through a filter (filter mesh not less than 0.2 mm). Pressure filtering is not permissible.

Devices and containers used in the processing of the honey must be made of materials legally permitted for the use with foodstuffs. Metal devices are to be of stainless steel.

Measurable Quality Criteria of the Honey

In addition to the legal requirements, the following are applicable: water content max. 18% (heathland honey 21.5%), HMF content in mg/kg max. 10, invertase units min. 10 (7 for acacia and linden honey) (analyses made according to AOAC (Association of Official Agricultural Chemists) standards).

Honey which does not fulfil the quality criteria in regard to HMF, enzyme or water content may only be marketed under the trade mark/ association name of BIOLAND as processing honey. No residue of chemical therapeutic agents may be traced in the honey which would indicate treatment of an impermissible nature.

Declaration

All stores and sales containers are to be marked. In addition to the legal details required, the BIOLAND business number and the contractual status are to be shown. The following marking is to be shown on the honey jars: As a result of the large radius of flight of the bees it cannot be expected that in all cases they will fly over only or mainly organically farmed areas (or in a similar form).

Pollen

The stripping facility should be arranged to avoid any injury to the bees. The pollen collecting basin should be arranged to remain sufficient pollen for the bees' own supply. The pollen within the pollen trap has to be protected against rain, moisture and direct sunlight. The pollen trap should be arranged to avoid the pollen to get lumpy (piling-up). For aeration the floor of the collecting basin should be equipped with a fine grid of special steel. The bottom of the hives are to be cleaned regularly. The pollen collecting basin must be of material legally permitted for the use with foodstuff and it has to be cleaned regularly upon need (but at least 2 times a week) carefully with boiling water or steam to avoid any mould.

Processing

At least once a day the pollen has to be removed and it must not be left within the pollen trap overnight. The pollen took must be dried immediately or frozen for a later processing. The drying air must not exceed 40° C degree of drying: The water contents must not exceed 6 %. The pollen has to be cleaned mechanically. Keep attention to that no foreign parts are in the pollen.

Packing and Storage

The pollen must be stored cool and dry. Storage and sales drums should be largely airtight to avoid humidity penetrating the pollen, and they have to protect the pollen against light. The storage drums are to state the year of harvest and the batch number. The sales packing has to indicate a batch number as well as the best-before-use date which should be limited to the end of the year following the year of harvest.

Further Bee Products

The use of the trade mark/association name of BIOLAND is possible for beeswax and beeswax products if the beeswax was originally produced from bees from a BIOLAND business.

In the processing of mead, the processing standards for the production of mead apply.

Conversion

During the conversion period, the hives, frames and combs are to be adapted according to the standards. They are to be marked accordingly. Available wooden hives with coats of harmless paint are regarded as being in accordance with the standards.

The BIOLAND wax cycle will also be introduced during the conversion. The use of the BIOLAND trade mark/association name is permissible for bee products from converted colonies, if these have been managed at least for period of one year in accordance with the standards. The converted colonies and their products are to be clearly marked. As long as no BIOLAND beeswax is available, the purchase of proven unpolluted wax from decapping and natural comb construction is permissible – also when using the trade mark for this particular colony – for the creation of central walls and the impregnation of the hives. Conversion is to be completed at the latest after five years.

This period may, if application to the BIOLAND association is approved, be extended. Stocks of honey from the period prior to conversion are to be clearly marked. The bee products can be marked with the trade mark/association name BIOLAND after the conversion period for the bee colonies is completed.

Fresh Water Fish Production

The general parts of the BIOLAND standards apply also to fish culture in as far as no exceptions are made in the following.

Types of Keeping

General Requirements for Keeping Fish

The fish may only be placed and reared in natural or almost natural waters such as earth basins and ponds. The use of plastic foils and keeping in nets is prohibited. The free movement of fish living in natural waters should not be hindered by the basin. A diversion ditch must be used in the case of new constructions or in reconstruction. Special regulations apply for propagation.

Retaining Fish

For retention purposes, ponds or basins with the smallest possible organic bed or suitable fish containers should be used. The period which the fish spend in the retention area is to be kept as short as possible.

Water Quality

Input water should fulfil the following minimum requirements:

- has no or only minor sewage water load
- has no harmful load from pesticides or fertilisers from farming
- has a sufficient oxygen content.

The quality of the water may not deteriorate significantly between input and exit points as a result of the fish culture. In order to evaluate water quality, the legally specified water quality classes will be applied. Aeration of the water is only permitted in exceptional circumstances to maintain life and not for the increase of growth.

Fish Culture and Care

Drying Out

When removing fish and subsequently drying out the pond or basin, appropriate damming measures must be taken to prevent sludge from being carried into the recipient.

Fertilising and Lime Fertilising

As fertilisers only organic fertilisers, as well as lime and stone powder are permissible. The use of quicklime for fertiliser purposes is prohibited.

Encroaching Water Plants

Encroaching water plants may only be removed by biological or mechanical means (e.g. cloudiness, joint weed). Chemical agents are not permissible. It is not permitted to burn off dams.

Biotope Unit

The business is obliged to maintain biotope structures, withdrawal possibilities and shelter for flora and fauna (guideline for total business is 5% of the pond area). At least 20% of the banks is to be left as a 1.5m wide sedimentation and reed zone.

Fish Stock Density

The stock of fish is to be oriented on the local conditions and the natural capacity of the pond.

The following maximum stock limitations apply:
- carp/ha: 3,000 K1 or 600 K^2.

In the case of mixed stock with tench and other non-predacious fish, the values are to be adapted in accordance with the weight of the fish. Stocking with predacious fish is to be in accordance with the natural feed content. Several types of fish are to be included in the stock.

Feeding

The basis for fish feeding is the natural feed content of the pond, by which the major part (more than 50 %) of the total feedstuff need of the production procedure must be covered. The pond's own production is to be used optimally by addition of feed of mainly plant origin. Additional feeding is to be carried out exclusively with feed from the business itself or from other farms of the BIOLAND Association or, if not available there, in accordance with the requirements of BIOLAND from other organically managed farms.

Treatment of Fish

Retention, transport, fishing and killing fish are to be carried out in such way that the fish are not subjected to any undue loads or stress. The fish may not be killed by means of suffocation.

Health of Fish

Permissible for treatment of fish are immersion baths with sodium chloride (common salt), quicklime or potassium permanganate. Furthermore the use of quicklime is permissible in case of immediate danger as treatment for the stock as well as after the occurrence of a disease as a hygienic measure to be spread on the wet floor of the pond or prior to the flooding of the pond. When prescription medication is used, the waiting period is to be doubled before the fish are put into circulation. All treatment measures effected are to be recorded in a treatment register.

Fish Reproduction and Breeding

The objective of breeding fish in fish culture is to have healthy, strong fish suitable for the location and which are to be found locally in the region. The use of hormones in reproduction is also prohibited.

Additional Purchase of Fish

In as far as such are available, fry fish must be purchased from other farms of the BIOLAND Association or, if not available there, in accordance with the requirements of BIOLAND from other organically managed farms. Fish purchased from conventional businesses must have spent at least two thirds of their lives in a BIOLAND business before they can be sold under the trade mark/ association name of BIOLAND.

Conversion

Adaptation of the fish culture to comply with the standards is carried out during the conversion period. At the beginning of the conversion, the water and the location are to be tested in regard to its suitability. Conversion, as a rule, takes place rapidly within two years, after a maximum of 5 years all production units must be taken into conversion. The trade mark/association name of BIOLAND may be used when the total production process (or, respectively, a total production unit) has been converted and the fish must have been kept at least 2/3 of their lifetime in compliance with the standards. When converting the complete business with all production branches in one step, it is allowed to use the trade mark/association name BIOLAND after a period of 24 months for all fishes available on the business at the time of start of conversion.

Horticulture and Permanent Crops

The general parts of these standards apply also to horticulture and permanent crops in as far as no exceptions are specified in the following. In farming without animals the supply of nitrogen must be effected as far as possible by leguminous growing in the business itself. The amounts of nitrogen fertiliser which is additionally required and permissible may be purchased in the form of external, organic additional fertilisers.

Vegetable Production

The total amount of fertiliser from the business and organic supplemental fertiliser to be used in free range vegetable gardening may not exceed 110 kg of nitrogen per ha and year. In greenhouses, carefully considering the problematic of nitrogen, the use of up to 330 kg of nitrogen per ha and year is permissible. In general, in vegetable gardening, Pt. 3.5.5 above is of particular importance. In order to control the nitrogen dynamics in the soil it is urgently recommended that N min. tests be carried out on a regular basis.

Soils and Substrates

Growing vegetables on stone wool, hydro-culture, nutritional film technology, thin layer culture and similar processes are not permissible neither are cultures in bags and containers.

Permissible is the growing of herbs in pots and similar products, whereas the container is sold together with the plant.

Hydragogue treatment for chicory is possible.

The use of peat to enrich the soil with organic substance is not permitted. It is also forbidden to use styrol mull and other synthetic materials in soils and in substrates.

Steaming Surfaces and Soil

Soil and substrates may be steamed. Flat steaming of the soil for the purpose of weed regulation is permissible.

Depth steaming to de-pollute the soil may only be permissible, if the plant protection problem may not be solved by other measures, e.g. change of crop, and requires express approval by the BIOLAND association.

Crop production under Glass and Foil

Heating greenhouses and foil covered premises must be within ecologically reasonable limits and, as a rule, should be limited to the reasonable extension of the culture period in autumn and to earlier starting in spring. In winter, the cultures should merely be kept free of frost (approx. 5°C). The young plant culture, forced sprouting and potted herb cultures are excepted. When choosing the system of heating and the fuel, the environmental compatibility should be taken into account. The buildings should be well insulated thermally. Used foils, fleece, etc. are to be recycled wherever possible.

Use of Technical Mulch Materials

A maximum of 5% of the free range area used for growing vegetables may be covered at any one time by mulch foil, mulch fleece or mulch paper. Businesses with less than 4 ha of area for vegetables may mulch up to 2,000 m² using the methods stated.

Harvesting and Preparation

When choosing the harvesting method and the date of harvesting and the preparation of the harvested products, the basic objective should be the achievement and the maintenance of an optimum quality for human nutrition.

Herb Cultivation

Preliminary Remarks

Medicinal and aromatic plants as a special group of cultures place higher demands on growing and processing. Their use, particularly in naturopathy, phytomedicine and cosmetics, necessitates detailed special knowledge in order to achieve the desired effectiveness of the active agents involved.

Advice on Production

In order to achieve the desired contents choice of location, fertilising, crop rotation and preparation should be adapted to comply as optimally as possible with the differing requirements of the individual species. The business should, therefore, obtain advice prior to entering into the field of growing medical and aromatic plants.

Selection of Location

As a result of the special significance of medicinal plants, the location is especially relevant. The minimum distance to roads should be 50 m and to field paths 5 m if no suitable protective planting has been effected.

Fertilising

In the year in which they are harvested it is not permitted to fertilise the cultures with liquid manure. Fresh manure may only be applied until the beginning of vegetation.

Preparation

In preparation the maintenance of an high quality is the prime principle. The devices used in processing must be designed in such way that the goods harvested are handled as gently as possible and no damaging substances (e.g. lubricants) can come into contact with the harvested goods.

Drying

The harvest for drug production must be taken into the drying plant immediately after processing. Materials detrimental to health such as PVC and treated chipboard may not be used. The drying room should form a closed unit. Direct heating with oil or wood or the extraction of moisture by means of chemical additives is forbidden. When drying, the temperature may not exceed the critical point at which a reduction in quality occurs. The drug must be dried to such an extent that its useful life is guaranteed (ideal figure is 8%). Different types of plants may not be dried together with one another when they have a negative effect on each other.

Further Processing and Packing

The main priority in further processing is the protection of the contents. For this reason, it is, therefore, inadvisable to mince or pulverise these too much. Further processing and packing of the drug should be effected as soon as possible after drying. Prior to packing, the drug should be cooled to room temperature. The packing material may not transfer any harmful materials to the drugs and must protect them from the effects of light.

Storage

The storage room must be protected against light, dry and as cool as possible. A weekly inspection of the goods in storage is mandatory to check for moisture, possible damage due to fungus or pests. Drugs of different types packed in permeable materials may not be stored on top of one another.

Shoots and Sprouts

In the production of shoots and sprouts the seeds, roots and rhizomes used must originate from BIOLAND propagation. If these are not available in sufficient quantities and qualities, then materials may purchased in accordance with the requirements of BIOLAND from other organically managed farms. Conventional sources are not permissible. The water used for the production of shoots and sprouts must be of drinking water quality. Any possible substrates and carrier materials used must be permissible and harmless in the sense of these standards. In cases of doubt, clarification should be obtained from BIOLAND.

Mushroom Production

Basic Principles

In addition to harvesting the mushrooms, the other important prodedures of mushroom growing (preparation of the substrate, inoculation, intermix growth phase) must take place in the business itself or in other businesses of the BIOLAND Association or, if not available there, in accordance with the requirements of BIOLAND in other organically managed farms. Substrates of foreign organic origin (intermixed or not) require the permission of the BIOLAND Association.

Substrate

The basic organic materials, substrate components and additives of the substrate (straw, cereal, bran, etc. and manure and compost) must originate from farms of the BIOLAND Association or, if not available there, in accordance with the requirements of BIOLAND from other organically managed farms. Only those sources of organic manure are allowed where it is guaranteed that only organic materials were used for the bedding of the animals. Should wood not be available from organic businesses in sufficient quantities, other sources are possible after careful testing. In order to obtain material which contains as few pollutants as possible, it must be possible to follow the origin of the wood in the chain of process in the selection of tree trunks, chips and sawdust; if necessary, their harmlessness should be proved by means of analysis. The use of peat as covering soil for growing champignon is permissible.

Disinfection and Plant Protection

Apart from composting only thermal processes are permissible for disinfecting the substrate. Appliances can be sanitised by means of alcohol or acetic acid.

The major objective in maintaining the health of the cultures is preventive plant control (hygiene, climatic conditions, mechanical protection against pests, etc.). The use of pyrethrum agents in mushroom production is not permissible.

Mushroom Brood

Attempts should be taken to obtain organic mushroom brood from other farms of the BIOLAND associations or in accordance with the requirements of BIOLAND from other organically managed farms. In the case of the business's own production of brood, the cereal used must originate from other businesses of the BIOLAND Association or, if not available there, in accordance with the requirements of BIOLAND from other organically managed farms.

Use of Energy

By the selection of suitable culture rooms, the energy used in the production of cultures must be kept as low as possible.

Fruit Growing

Basic Principles

Fruit growing, as an intensive permanent culture, places special demands on the design of the total business. Prerequisites for successful, organic-biological fruit production are:

- the selection of suitable varieties, (under) stock and forms of training,
- the generation and maintenance of an ecological balance between pests and beneficial animals,
- creation of a favourable microclimate in the fruit plantation, and
- the use of measures which strengthen the health of the plants and prevent illnesses and pests.

Fertilising

The total quantity of nitrogenous fertiliser used may not exceed 90 kg N per ha of fruit plantation and year. In businesses in which there are no animals, this amount may be purchased.

Supporting Material

Tropical or sub-tropical woods may not be used as supporting material. The tropical grasses, bamboo and tonkin are permissible.

Viticulture—Soil Care, Greening and Fertilising

In order to reduce the problems and disadvantages of the mono-culture in vineyards and in the process of extensive growing to ensure the production of grapes, juice and wine of a high quality, the yield giving vineyard must be greened throughout the year. Greening is to be regulated by mechanical means in such a way that a mixture of various plant species is maintained and beneficial animals are attracted by the blossoming flowers. For special soil care measures, in dry periods in summer and in care of young plantations, the greening can be turned over in part. If the soil is kept open for more than three months, a soil covering of organic material must be applied. Re-sowing must be with a well mixed variety among which there must be a considerable part of leguminous plants. The nitrogen balance should be considered when doing this. In the case of vineyards on steep slopes with skeleton rich soil, all measures should be carried out according to the local conditions. Changes in the complete surface greening throughout the year are to be recorded on the inspection sheet. In wine growing, the nitrogen fertilising should not exceed a total volume of 150 kg N/ha in a three-year cycle whereby the fertiliser available for the plants may not exceed 70 kg N/ha in any one year.

Supporting Material

Tropical or sub-tropical woods may not be used as supporting material.

Plant Protection

In the sense of preventive plant care all measures adopted in vineyard cultures are to be effected in such a way that the resistance of the vine is increased, the amount of damage by infectious agents reduced and useful organisms supported. It is, therefore, essential to select vine varieties, vine cultivation and stock formation, foliage work, vine nutrition and soil care suitable for the location.

In the case of plant protection measures from the air extending beyond the business itself, the whole of the operation is nevertheless subject to the standards described here. It is to be agreed in writing with the BIOLAND Association which plots of land can be regarded as being free of pesticides and drift from helicopter spraying. Extent, form and location are to be taken into consideration. Grapes from this lands as well as the products prepared from them like wine and juice must not be marketed under the trade mark/association name of BIOLAND.

Ecological Niches

Every vintner is obliged to plant and tend reasonably a part of its vineyard area as an ecological niche. Efforts should be made to reach at least one percent of the vineyard area. The ecological niches must, in order to interrupt the mono-culture, be distributed throughout the area. The situation of the surroundings and communal measures for nature conservation have to be considered.

Hop Cultivation

Location and Area

When the location necessitates this, protective plantings must be created latest within five years after start of conversion (when directly adjactant to conventional areas) or, respectively, ecological compensation areas (in cleared areas). New hop cultivation must be a border field or a separate area. In order to prevent the immission of conventional plant protection agents, the distance to conventional hop cultivation areas must be at least 10 m. Where this is not possible, the outer rows must be plucked separately and marketed conventionally.

Support Material

Wood as support material for new hop cultivation units must be from species of trees growing in the country in which the business is located. Impregnation may only be carried out with agents which exhibit high environmental compatibility.

Greening

Greening of hop cultivation areas throughout the whole of the year is to be effected using mixtures of grasses, herbs and leguminous plants of appropriate species. In order to prevent nutrients being washed out, greening is mandatory, at least from the time of harvesting until spring.

Fertilising

The nutritional supply of the hops must be mainly in the form of fertilisers generated by the business itself and a balanced green fertilising is to be effected. The total amount of the fertiliser from the business itself and external organic complementary fertilisers used (compare 10.1) may not exceed 70kg of nitrogen per ha and year.

Preparation

The use of sulphur for conservation is prohibited in drying and processing.

Records

The operator agrees to keep a record in which all fertilising, plant protection and green fertilising measures are noted documenting the amounts used and the date for each hop cultivation area. The hop cultivation involved must be noted on the weighing slip with official sealing.

Ornamental Plants, Herbaceous Plants and Woody Plants

Fertilising and Soil Care

The use of nitrogenous fertilisers on free range culture areas in which tree nursery cultures are cultivated is limited to 90 kg N/ha and year, otherwise limited 110 kg N/ha and year. It is urgently recommended that annual mineral nitrogen content (Nmin-method) checks are carried out annually to control the nitrogenous dynamics of the soil. For areas which will probably remain uncultivated for more then 12 weeks during the vegetation period and, as far as possible, also throughout the winter, green fertilising is to be carried out.

Surface Sealing

Sealing free range storage areas for pots and containers is only permissible for the purpose of reusing water.

Plant Health and Regulation of Weeds

In businesses operating as ornamental plant, herbaceous plant and tree nurseries measures for preventive plant protection are of central importance. This includes, among other things, the choice of suitable resistant types, the selection of healthy seeds and plants, optimum culture processing with appropriate plant density, adapted crop rotation, fertilising and management of humus. Measures must be taken in the business to further the self-regulatory powers of the ecological system.

The regulation of weeds is effected in accordance with 3.8. Flat steaming is permissible in greenhouses to combat weeds. In-depth-steaming and steaming of free range areas is only permissible, if the plant protection problem may not be solved by other measures, e.g. change of crop, and requires express approval by the BIOLAND association.

Seedlings

If no organically reared seedlings are available recourse can be made to conventional sources following approval being issued by the BIOLAND Association. These conventional seedlings must pass through conversion in special areas. Should they be sold prior to completion of conversion, they may not be designated as being organic. Use of the BIOLAND trade mark/association name is prohibited in such cases.

Additional Purchase and Trade Goods

If conventionally finished products are purchased this must be clearly recognisable in the business at all times (purchase, insertion, further culture, etc.). This is to be ensured by means of suitable measures (e.g. labelling, separate beds or patches). In relation to turnover of the plant products sold, the majority must originate from ecological production.

Soils and Substrates

Wherever possible, peat should not be used. The peat content of substrates may not exceed a maximum of 50 vol. % in the case of tree, herbaceous and ornamental plant cultures and 80 vol. % in the case of seedlings. In the case of plants which require a low pH value for their growth, this ruling can be deviated from. Purchased composts, peat substitutes and additives must be examined in regard to their environmental compatibility and, in particular, to their pollutant content. Synthetic additives (e.g. styrol mull, hygro mull) and stone wool are not permissible. Soils and substrates may not be steam treated.

Containers for Cultures

Attempts should be made wherever possible to use containers of decomposable materials (e.g. recycled paper, wood fibres, flax, jute, hemp) or earthenware containers. Pots and bowls of plastic must be of a stable material and can be reused. The material must also be recyclable. Containers made of PVC are not permitted. Available pots which do not meet these criteria may be used up during the conversion period.

Storage

BIOLAND products must be stored in such way that the quality is not negatively affected by storage. The treatment of the harvested products with chemical storage protection agents (insecticides, fungicides or similar) and storage in containers made of materials with substances which may be detrimental to health, washing stored fruits with chemical cleaning agents, further ripening with chemical substances, the use of germination prevention agents and radioactive irradiation are forbidden. Cleaning of the storage facilities is to be effected using measures which exclude placing environmental loads on the goods stored.

Objectives of Processing Standards

Processors of BIOLAND products continue the efforts of organic agriculture to maintain the natural living conditions for plants, animals and human beings on a long-term basis. BIOLAND products produced in accordance with these standards are characterised by their high quality in taste and their high values in health, ecology and culture. The processing standards, in the sense of high nutrition, are designed to ensure a "Vollwert" nutritional, physiological and ecological quality standard of the final products while taking social tolerance of trade and processing steps into consideration. A further objective of these standards is the creation of the greatest possible degree of transparency, in particular for the consumer.

Scope of Validity of the Processing Standards

All BIOLAND processors, production businesses with their own farm processing and commissioned businesses are obligated to comply with these standards. Processors in the sense of these standards are natural persons and legal entities who/which, by means of cleaning, treating or processing or filling BIOLAND products, achieve an added value and who/which have concluded a contract with BIOLAND e.V. for the use of the trade mark. In addition to the general processing standards, the product-specific standards which are regulated either in the contract or in the branch standards also apply. The appropriate branch standards contain, in particular, regulations concerning scope of validity, additives and processing aids, processing methods, packing, hygiene, declarations and quality assurance.

Ingredients and Processing Aids

Basically, only ingredients from BIOLAND production are

permissible for BIOLAND products being processed. They are to be acquired from producers and processing businesses which are connected to BIOLAND by means of a producer's or processing contract, respectively. The use of foreign ingredients from organic production in BIOLAND processed goods is possible in founded exceptional cases to a limited extend, if these ingredients:

- are not produced in BIOLAND producers or processing businesses
- are evidently not produced in sufficient quantity and/or quality available from BIOLAND producers and processing businesses.

Prior to the use of such foreign ingredients from organic production, the processor has to file a formal application to BIOLAND for express approval, except BIOLAND has issued a general approval for special goods or group of goods (e.g. seed, spices, exotic fruits) and has informed processors about this. An express approval is always limited in time and quantity. Prerequisite for the use of foreign ingredients from organic production is that these ingredients are recognised by BIOLAND. In the approval of foreign ingredients BIOLAND observes the following priority:

- ingredients or goods from businesses that are certified by IFOAM accredited organisations or from other organisations recognised by BIOLAND 2. ingredients or goods from businesses that manage at least according to EU regulation 2092/91.

In principle the use of ingredients from conventional production is not permissible. If it is proven that certain ingredients from organic production are not available, conventional ingredients may be used in exceptional cases to a part of a maximum of 5 % as far as these are listed in the EU regulation 2092/91, annex VI, part C. A BIOLAND product may not contain the same ingredient from organic and non-organic origin.

Further Additives and Processing Aids

Only additives and processing aids which cause no damage to health may be used. Water and salt may be used as ingredients in the production of BIOLAND products and are not included in the percentage calculations of organic ingredients. Any additives and processing aids permissible for the production of BIOLAND products are itemised as positive listings in the product-specific BIOLAND processing standards. If there are no regulations for certain products, annex VI section A and B of the EU regulation 2092/91 as well as annex 4 of the IFOAM Basic Standards are authoritative. BIOLAND products may not be enriched with minerals (including trace elements), vitamins, amino acids or similar isolated substances, except the use in food is legally prescribed and approved

Processing

Processes are to be used in the treatment and processing of raw materials which – in accordance with the latest status of scientific knowledge – maintain the ingredients of the foodstuffs in an optimum manner and in the sense of wholefood nutrition. This has to be ensured by applying processing methods and techniques the basis of which are biological, physical and mechanical nature. Extraction shall only take place with water, ethanol, plant oil, carbon dioxide and nitrogen, These shall be of a quality appropriate for their purpose.

The processes must ensure the most economical use of resources such as water, air and energy sources. The appropriate branch standards contain recommendations for processing methods anddevices. The processor has to take all required measures-to ensure the identification of BIOLAND products or parts by clear labelling of the product itself as well as of packing, cases, means of transport, shipping documents etc., to prevent co-mingling, contamination or confusion of BIOLAND products with Non-BIOLAND products,-to prevent the contamination of BIOLAND products by pollutants and residual matters, including impurity by cleaning and decontamination; if necessary, the production rooms and facilities are to be purified and disinfected thoroughly. The processor has to take care that these measures securing the quality are executed also in the previous processing stage, including the subcontracted work. Especially all businesses processing, storing or transporting also conventional products apart from BIOLAND products have to carefully and completely purify the means of transport, storage rooms and receptacles (silos), facilities, equipment or appliances before taking any BIOLAND products.

Direct or indirect contact with non-permissible substances (e.g. pesticides) and BIOLAND products while doing pest control measures has to be avoided at all times. In case any non-permissible substances or methods have been applied directly on foodstuff or stocks, the products in concern may not be marketed as BIOLAND products. The processor has to take all necessary safety measures to avoid a contamination, including the removal of BIOLAND products from the store or the processing facility. The application of non-permissible substances on facilities or appliances may not contaminate the BIOLAND products produced therein or therewith. In case of doubt the processor has to analyse such products on residue loads. The measures for pest control authorised in the BIOLAND contractual businesses are listed in the BIOLAND standards for pest control in storage rooms and operational premises.

Packing Materials

The choice of packing materials is made in accordance with the following criteria:

- packaging materials must be unharmful physiologically, especially with respect to the migration of health-hazardous substances into the food, and as environmentally friendly in production as possible.

- No packing materials as well as storage rooms, silos or other storage tanks may be used which contain synthetic fungicides, food preservatives or vermin destruction agents. BIOLAND products may not packed in used bags or cases which came into contact with substances possibly influencing the intactness of BIOLAND products or their ingredients.

- The packaging volume must be reduced to the minimum amount technically required. Hereby, ecological requirements are to take priority over marketing technical and costing aspects.

- The packing materials should be recyclable within the scope of refuse reprocessing. · Plastics which are difficult to decompose (for example, such as PVC) or, respectively, plastics which are manufactured in a manner which causes an irresponsible load to be placed on the environment may not be used.

- Aluminium foils or foils with aluminium content or combined packaging may only be used following express approval by BIOLAND e.V. The processor is obligated to attempt to find alternative forms of packing.

- Non-returnable packing may not be used if returnable packaging is possible and feasible. · The appropriate branch standards contain recommendations/positive listings for packaging materials.

Labelling of Processed BIOLAND Products

When designing the packaging, the "Standards for the Design of Packaging for BIOLAND Products" as currently amended must be complied with in order to present the consumer with an easily recognisable BIOLAND total assortment. Labelling and declaration must be in accordance with the requirements of the Foodstuffs and Consumer Goods Act (LMBG). Ingredients and additives to BIOLAND products are to be declared fully and all ingredients in BIOLAND products must be declared in full extent and-in the case of multi ingredient products-listed in the sequence of their weight percentage. Herbs and spices may be listed within a collective expression, if their percentage is less than 2 % of the total weight of the product. It has to be stated clearly which ingredients are from organic origin and which are not. If additives are used they have to be listed with their product name or their original name. A class or group designation of the additives is not permissible.

Storage and Transport

General conditions in regard to this are to be found in Section 6 of the BIOLAND Standards. BIOLAND-and non-BIOLAND-products may not be stored or transported together except the BIOLAND-products are clearly labelled and separated physically. A control system for storage conditions including controlled atmosphere, temperature control, drying and moisture control is allowed. Further details are specified in the branch standards.

Transparency and Product Identification

Retention Samples

The processor is obligated to draw a sample from each batch of raw materials delivered, to mark these with the date of delivery and the name of the supplier. In addition, samples from the finished resp. half-finished products must be drawn. These retention samples are to be kept until expiry of the „best before" date of the processed goods resp., in case the indication of a „best before" date is not necessary, for an appropriate period. Exceptions can be made in individual cases for certain products or processing areas (e.g. in the case of easily perishable raw materials) in the appropriate branch standards if the aforementioned obligation to draw and keep samples is not economically justifiable or practically feasible.

Raw Material Identification

Each processor is obligated to ensure by means of suitable measures within the scope of the quality control procedures of the business that the BIOLAND raw materials supplier can be identified at all times.

Execution and Inspection

Responsibility in the BIOLAND Association

The basic concepts and the major contents of the general processing standards and the branch specific standards are passed by resolution at the Federal Delegates' Assembly.

The Advisory Council responsible for the processing standards or, respectively, its subcommission, in which representatives of the processors under contract from individual product areas also act in an advisory role, develops and continuously monitors the branch-specific processing standards. It is also the task of the Advisory Council to reach decisions on alterations and extensions of these standards and to make recommendations in regard to them. The Federal Board can then reach a decision on the alterations to the standards unless the objectives and the contents of the standards are affected to such an extent that the Federal Delegates' Assembly is required to reach a resolution in regard to this.

Alterations to Products Being Processed

Each processor is obligated to inform BIOLAND e.V. in a timely manner of any major alteration in the processing, the additives, the packing or the design of his products within the scope of the processing or, respectively, the design standards.

New products or planned alterations to existing products being processed which cannot be brought into line immediately with the requirements of the general and branch-specific processing standards must be approved by BIOLAND e.V. An application is to be submitted to the appropriate Advisory Council which will discuss and reach a decision on the application. If required, the processor will supply information on all of the ingredients of the product and the methods of processing. Should differences of opinion arise, an attempt will be undertaken with the processor to reach an acceptable solution on the basis of the processing standards. If this is not possible, the Federal Board will decide on the way of action.

Inspection

Each processor is regularly inspected in regard to compliance with general and branch-specific processing standards. The processor is obligated to place the necessary documents and records at the disposal of the person from the inspection body authorised by BIOLAND e.V. to carry out the inspection.

The latter are bound to maintain secrecy in respect to third parties. In the case of a founded suspicion, BIOLAND e.V. is entitled to inspect the business during normal working hours without giving prior notice. The processor will place the inspection results according to EU regulation 2092/91 at the disposal of BIOLAND e.V., so that the BIOLAND inspections can be based on them.

Contamination Tests

As a result of the general loads be placed on the environment or other possible sources pollutants can also find their way into BIOLAND products. The processors are therefore obliged to carefully analyse and determine the weakest points resp. risky areas for potential pollutants of the products. Based on this, a programme according to the HACCP concept for systematic pollution test of BIOLAND products has to be established.

The pollution analyses must be executed through acknowledged testing laboratories based on latest techniques with regard to sample taking, test extent, analysis programme and analysis method. The results of the pollution analyses are to be recorded and made available upon request to BIOLAND as well as the responsible inspection body.

Obligation to inform and to Register

The processor is obliged, over and above the legal obligation to inform according to the Foodstuffs and Consumer Goods Acts (LMBG), to immediately inform BIOLAND in case of any assumption or doubt that raw materials, ingredients or BIOLAND products being processed do not correspond to the regulations serving for the protection of human health, or if they cannot be put into circulation for any other reason.

Marketing

Basic Principles

Marketing is carried out in close co-operation with BIOLAND e.V. in order to ensure that the quantitative and qualitative requirements of the market are considered. The products are to be brought to the consumer by the most direct means possible. Marketing must be so transparent that the consumer can follow the path of the product from the producer through to the consumer. Only marketing activities (in particular in regard to advertising/sales promotion, the choice of distribution method, price and product design) may be adopted which do not contradict the objectives and measures of BIOLAND e.V.

Production Recording

The contractual business is obligated to participate in the annual production recording (business reports).

Marking and Packing

Contractual businesses are obligated to mark their products, produced in accordance with the standards, with the trade mark BIOLAND, their address and business number. BIOLAND e.V. designs marking and packing material. The use of other marking or packing belonging to the business itself requires the express approval of the BIOLAND State Association. Uncontrolled packing material may not be used.

Additional Purchase

Trade goods destined for direct marketing are to purchase from BIOLAND contractual businesses preferably.

Bulk goods purchased which do not originate from BIOLAND businesses must be clearly marked as such in marketing by stating of the farmers' association or the certifying organisation respectively. The purchase of conventional goods for direct marketing is not permissible. Products which are not supplied in organic quality are exempted from this, but necessitates exemption approval by the appropriate BIOLAND State Association.

Sales to Commercial Buyers

In selling to commercial buyers, the contractual partners of BIOLAND e.V. or, respectively, other trade partners with whom BIOLAND e.V. co-operates, are to be given preference.

Use of the BIOLAND Trade Mark

Contractual businesses are obligated to actively and continually promote and care for the BIOLAND trade mark. All activities are to be aimed at increasing the degree of awareness of the trade mark and clearly marking and preventing the misuse of the BIOLAND products on the markets supplied. The businesses will inform BIOLAND e.V. immediately of any misuse or unauthorised use of the BIOLAND trade mark on the part of association members or other users of the trade mark on the market and in advertising.

Commercially Operated Farm Shops and Market Stands

The standards apply also to all of the non-agricultural parts of the business associated with the business itself such as farm shops, market stands in as far as these appear for the consumer to be connected to the business.

Contractual and Inspection Measures Responsible Bodies

The responsibilities for all matters in connection with these standards and for the rights and duties of the members are regulated in the articles of incorporation (statutes) of BIOLAND Verband für organisch-biologischen Landbau e.V. (Federal Association).

Conversion—Producer Contract

The sale of products under the trade mark/association name of BIOLAND presupposes the conclusion of a producer contract with the issue of a business operating number which carries the obligation to comply with the standards of BIOLAND e.V. Producer contracts are concluded in relation to areas and to single persons. Prerequisite for the conclusion of a contract is membership in BIOLAND e.V. When a contract is issued, a visit is made to the business by a person authorised by BIOLAND. Each producer contract will be accompanied by a binding conversion plan. All of the conversion steps will be specified in this and, in particular, the resulting possible commencement of the use of the trade mark/association name of BIOLAND for the individual branches of the business. Any subsequent deviating agreements between the business and BIOLAND must, in order to achieve validity, be made in writing. In the case of pending difficulties in plant or animal production or in the marketing or in the case of factual uncertainty the manager of the business must contact BIOLAND in due time prior to reaching a decision (as a rule, in writing).

Conversion of Total Business

Contractual businesses are obligated to cultivate all lands and production branches of the business in accordance with the standards as currently amended. The keeping of utilizable animal species, for which these standards do not provide express regulations, requires the approval of BIOLAND, likewise the use of the trade mark/association name of BIOLAND for the products of such branches of production.

Use of Trade Mark for Plant Products

The use of the trade mark/association name of BIOLAND with the addition of "from conversion" can be used for plant products consisting of a single ingredient of an agricultural source when the area has been cultivated in accordance with the standards for 12 months prior to the harvest. For reasons of importance, this period can be extended. The trade mark/association name of BIOLAND can be used if the land is cultivated in accordance with the standards for a period of 12 months prior to sowing and in the case of perennial cultures for 36 months prior to harvesting. If new areas (fields) are added to the business, then these must be put through the process of conversion. Efforts should be made also in the cases of rented areas, to achieve long-term, organic-biological cultivation. It is not permitted to simultane usly plant the same types of plants on different areas of the business which are at different stages within the conversion process. Exceptions to this are:

- perennial cultures
- growing of vegetables and ornamental plants when the cultures which are planted parallel to one another are easily differentiated from each other
- growing of fodder plants.

Use of Trade Mark for Animal Products

Animal products may first be marked with the trade mark/ association name of BIOLAND at the earliest when the beginning of the conversion of the areas for fodder/feeding took place at least 12 months prior to this and the subsequent following periods for conversion for feeding and keeping of all of the animal species have been adhered to in accordance with the standards:

- Eggs: 6 weeks
- Milk: 3 months (from 24.08.2003: 6 months)
- Cattle: 12 month & in any case minimum three quarters of the animal's lifetime;

The use of the trademark/association's name "BIOLAND" is not permitted for beef from cattle that was born on conventional farms and/or raised with feedstuffs non-compliant with these standards.

- Sheep & Goats: 6 months
- Pigs: 4 months (from 24.08.2003: 6 months)
- poultry meat 10 weeks (if put in barn no later then the 3rd day of life)
- Fallow-Deer and Red-Deer: 12 month

- Rabbits: The use of the trademark/association's name "BIOLAND" is only permitted if the animals were kept and raised with feedstuffs compliant with these standards from birth.

Feed/fodder in accordance with the standards is specified as being:

- organically produced fodder: fodder from lands which has been managed organically a minimum of 24 month prior to sowing, in case of permanent grassland 24 months prior to the beginning of the use as fodder.
- feed/fodder permissible in accordance with 4.4.2.

The use of the trade mark/association name of BIOLAND can start earliest when the all animals of the same species have been converted. In beekeeping, the use of the trade mark/association name of BIOLAND can be used at the earliest 12 months after the commencement of conversion if the bee colonies comply with the requirements of 4.10. In fish culture, the use of the trade mark/ association name of BIOLAND can be used at the earliest 12 months after the commencement of conversion if the ponds comply with the requirements of 4.11. Keeping of poultry in cages in the business must be discontinued prior to any use of the trade mark.

Simultaneous Conversion of the Total Business

In case of simultaneous conversion of the total business (i.e. all lands and animal categories) all animal products produced from animals, present at the commencement of the conversion, and its progeny can be marketed under the use of the trade mark/association name BIOLAND in deviation from 9.2.4.1, provided that the animals are fed mainly with the business' own fodder.

The use of the trademark/association's name "BIOLAND" is not permitted for beef from cattle that was born on conventional farms and/or raised with feedstuffs non-compliant with these standards.

Conversion Deadlines

Conversion is carried out without delay, in plant culture in one step. In exceptional cases this can be effected in steps and must be completed at the latest after a maximum of 5 years. 9.2.6 Non-permissible Operating Resources Resources, the use of which is excluded by the standards, may no longer be available in the business.

Further Training

The managers of businesses must possess the necessary theoretical and practical skills. Minimum evidence of this, in addition to the prior completion of agricultural training or professional experience, is supplied by successful attendance of an introductory course in organic biological farming. The exchange of experience and the discussion on the operating conditions are important basic factors of further training and the gaining of the necessary confidence. Each manager is a member of a regional or specialised group. The business managers participate as actively as possible in group work and in the exchange of experience in the group.

Inspection

The BIOLAND Association will check compliance with its standards by the contracting businesses (producers). Inspection checks will assist the contracting parties in the further development of the business in the sense of these standards.

Inspection Procedure

The inspection of contractual businesses is composed of supplying written answers to a questionnaire (operating business report) and an inspection visit for which an inspection report will be written. It will be carried out at least once per year by an inspector authorised by the BIOLAND Association who is both independent and competent. The business inspected in this manner will receive a copy of its business operation report and the inspection report. In the case of a step by step conversion, the inspection of the business will also include those parts of the business not yet converted. A commission for recognition set up by BIOLAND for this purpose will decide annually on instructions, warnings and sanctions. The basis for any decisions of this nature is a catalogue of sanctions published by the BIOLAND Association.

Necessary Documentation and Information from the Business

The businesses must keep clear records of all points to which these standards apply: cultivated area, crop rotation, fertilising, plant protection, animal stock, keeping, feeding, treatment of animals, marketing, storage and purchasing from external sources. The BIOLAND Association is entitled to require the member to supply and to store data which will serve to record production quantities and for inspection purposes. Additions to the area must be reported to the

Association without delay. This applies also to any change in the business address or change in the management of the business. The BIOLAND Association can require the business to supply soil examinations, quality tests and examinations of residues. If there is evidence of a breach of the standards, the costs of the examinations will be borne by the businesses.

Right to Examine Records and Right of Access

The business is required to afford the representative of BIOLAND access to the whole of the business in order to carry out inspections. The BIOLAND Association is entitled at any time to have the business and the books of the member examined by an employee or an authorised person. Such person is sworn to secrecy and may not pass on any information to any third party.

Commencement of Validity and Transitional Arrangements

Amendments to the standards become valid on their being published in the association's organ, the magazine bio-land. The standards amended by resolution of the Federal Delegates Assembly as of May 3, 2000 are considered as being published as of August 24, 2000. Businesses which at the time of the appropriate amendment to the standards have concluded a producer contract with the BIOLAND Association and do not yet fulfil the amended standards have, with effect from the date of publication, one year's time, in case of constructional changes in buildings two years –unless another deadline is expressly determined-in which to adapt to comply with the new conditions with proviso to further going conditions of the EU-regulation 2092/91. There are no transitional periods for the construction of new barn buildings.

Limited to the following rulings the above mentioned transitional periods can be extended until 2010 for buildings existing before 24.08.1999 (in case of buildings with animal tethering systems existing before 24.08.2000), if it has been approved by the BIOLAND Association:

- space requirements for interior and exterior areas of barn buildings for mammals
- tethering systems for cattle
- space requirements for the open run for poultry
- maximum size of poultry barns
- total useable area of the barns for fattening poultry
- combined length of the fly-out-openings of poultry barns

Fertilisers and Soil Conditioner from Organic Operations:
- stable/shed manure and poultry manure
- liquid manure following processing
- liquid manure
- composts from organic refuse
- substrates from cultures of mushrooms
- straw.

Fertilisers from Conventional Operations:
- cattle manure
- sheep and goat manure
- horse manure.

Organic Complementing Fertilisers and Soil Conditioner Agents as well as components of substrates:
- quality assured composted organic household refuse (Bio-compost) and plant composts (composts from greens) according to the amended BIOLAND regulations
- quality assured composts from bark of chemically untreated wood after cutting
- saw dust, wood cuttings and wood ashes of chemically untreated wood after cutting
- peat, only in substrates and with the restrictions
- the following products and residue of animal origin: horn shavings, horn meal, feather meal, hair meal and bristles.
- products and residue of plant origin (e.g. castor-oil groats, rape groats)
- vinasse (only in plant gardening and permanent cultures)
- algae and algae products

Mineral Complementing Fertilisers:
- mineral powder
- clay
- raw phosphate (ground, soft texture, not partially processed)
- thomas phosphate
- raw potassium salt (e.g. Kainit)
- patent potassium (potassium magnesia)
- potassium sulphate

- magnesium sulphate
- magnesium carbonate
- Calcium carbonate, dolomite lime, shell lime, marine algae lime
- basic slug, converter lime, smelter lime
- gypsum of natural origin
- calcium chloride
- carbo-lime from the processing of organically cultivated sugar beet)
- elementary sulphur
- trace element fertilizers.

Preparations:

- Preparations from micro-organisms to use in soils, composts and substrates, e.g. for advancing resetting processes if their compositions comply with these standards.

Permissible Plant Treatment Agents and Methods

For the use of plant protection and plant care agents the legal regulations, above all the conditions of the EU-regulation 2092/91 and the German Plant Protection Act (Pflanzenschutzgesetz), have to be observed. Only the restrictions in the use exceeding these regulation are listed below.

Biological and Biotechnical Measures

- planned use of beneficial animals (e.g. predatory mites, parasitic hymenopter)
- insect traps (glue traps)
- culture protection nets, mulch foils etc.
- use of pheromones.

Plant Protection and Care Agents

The agents specified may only be used in as far as these are not used in combination with other plant protection agents which are not named here.

Generally Permissible Agents::

- stone meal
- bentonite and prepared aluminium oxide
- algae meal and algae preparations

- water glass (sodium silicate)
- herb extracts, herb liquid manure and teas (e.g. nettle, horsetail, onion, horse radish, parsley fern)
- Azadirachtine from Azadirachta indica (Neem tree)
- quassia from quassia armara
- mineral oils and paraffin oils
- plant oils
- ethyl alcohol
- potassium soap
- iron-III-phosphate
- milk and whey products
- micro-organisms (bacteria, virus, fungi), e.g. bacillus thuringiensis preparations
- sodium hydrogen carbonate
- lecithin
- quartz (siliciumdioxide)
- pyrethrines from Chrysanthemum cinerariaefolium
- wettable sulphur
- sulphuric lime (calciumpolysulfide)
- potassium permanganate
- copper preparations (max. copper volume 3 kg/ha and year, in hop cultivation max. 4 kg/ha and year, in potato cultivation only with permission of the Association. If agents with copper content are used, the copper content of the soil must be continuously monitored by means of soil examination).
- hydrolysed protein (enticing agent).

Design and Optimisation of Animal Breeding

There are two fundamental questions faced by animal breeders. The first asks: "What is the best animal?" Is the best Labrador the one with show-winning conformation or the one with exceptional retrieving instinct? Is the best dairy cow the one that gives the most milk; the one with the best feet, legs and udder support; or the one that combines performance in these traits in some optimal way? These are matters of intense debate among breeders, and, in truth, no one has all the answers. The question is an important one, however, because the answers determine the desired direction of genetic change for breeding organisations and people keeping farm or companion animals. The second question asks, "How do you breed animals so that their descendants will be, if not "best", at least better than today's animals?". In other words, how can we genetically improve animal populations? This question involves genetic principles and animal breeding technology, and is the subject of this course.

What is the Best Animal

"Best" is a relative term. There is no best animal for all situations. The kind of animal that works best in one environment may be quite different from the best animal under another set of circumstances. When we describe animals, we usually characterise them either in terms of appearance or performance or some combination of both. In any case, we talk about traits. A trait is any observable or measurable characteristic of an animal.

Some examples of *observable* traits –traits we would normally mention in describing the appearance of an animal-are coat colour, size, muscling, leg set, udder conformation, and so on. Some examples of *measurable* traits –traits we would likely refer to in describing how an animal has performed-are body weight, daily milk production, time to run a mile, etc. There are hundreds of traits of interest in domesticated animals. Note that in none of the examples of traits mentioned above is the appearance or performance of a particular animal described. An animal may be red and weigh 343 kilograms at 1 years of age, but *red* coat colour and *343 kg* yearling weight are not the traits-the traits are simply coat colour and weaning weight. *Red* and *343 kg* are the observed categories or measured levels of performance for the traits of coat colour and yearling weight.

They are the phenotypes for these traits. In animal breeding, we are mainly concerned with changing animal populations genetically. From a genetic point of view, therefore, we want to know not only the most desirable phenotypes, but the most desirable genotypes as well. That is because an animal's genotype provides the genetic background for its phenotypes and it is the genetic material that is passed on from parents to its offspring. Summarised in an equation : where P represents an individual's phenotype, G represents its genotype, and E represents the environmental effects-the effects that external (nongenetic) factors have on an animal's performance.

In other words, its genotype and the environment it experiences determine an animal's phenotype. The word *genotype* is used in several ways. We can speak of an animal's genotype in general, referring to all the genes and gene combinations that affect the array of traits of interest to us. An example used later on in this section involves a "tropically adapted" genotype. In this case, the genotype includes all the genes and gene combinations affecting heat resistance, parasite resistance, and other traits that make up tropical adaptation.

This sense of the word *genotype* is generally implied in this chapter. We can also speak of an animal's genotype for a particular trait, referring to just those genes and gene combinations that affect that trait (e.g., heat resistance). Or, as we will see later in this course, we can limit the definition of genotype even further in which case it refers to a particular gene only (e.g., an animal has genotype AA for the kappa-casein gene). In any case, the genotypes of our animals' descendants are what we can change with breeding methods. Favourable changes in genotypes result in improved phenotypes.

To answer the question "What is the best animal?" we need to determine what traits are of primary importance and what genotypes are most desirable for those traits. Most breeders, if they have some experience, have an opinion about the key traits and better genotypes. A Thoroughbred breeder, for example, might describe the perfect animal as ".... fast, but with enough endurance and heart for the longer distances, and easily rated". A pig breeder version might be ".... a healthy pig with a good growth and good carcass quality." There are probably as many opinions of this sort as there are breeders and for the most part they are quite subjective.

In order to develop a sense of the important traits and best genotypes in a more objective way it is important to understand the role of the genotype in the system of the farm. This means that the importance of traits will depend on the physical environment under which animals are kept, the management system as well as economic factors. If you think about it, it will become clear that a number of the components of the system will interact with each other. For example, the best preventive health program (management) depends on the kinds of pathogens in the area (physical environment) and the costs of vaccines, dewormers, etc. (economics). To determine which health program is the most cost-effective, you must have knowledge of alternative programs, local pathogens, and treatment costs and understand how treatment programs interact with these other factors to affect profitability. Similarly, the best genotype depends on the local environment, the management practises in use, and the costs of inputs and prices of animal products. To determine the best genotype, you must have knowledge of environmental, management, and economic components and understand how they interact with the genotype to affect profitability. Knowledge of the function of the animal and the interactions between the genotype and other components of the system is necessary if we want to develop sensible goals for breeding programs, in other words, if we want to develop appropriate breeding objectives. Knowing, for 1 This mathematical expression is oversimplified but it will do fine for the purposes of this discussion. Later on we will see that there might also be an interaction between the G and E. The genotype of domestic animals determines the degree to which the animals are suited for their function in society. The key to determining the traits of importance and optimal genotypes for those traits is a thorough analysis of the function of the animal in the entire system and an understanding of the many interactions among components of the system example, that parasite resistance is critically important in tropical climates, breeding objectives in the Tropics emphasise traits such as tick count (a measure of tick resistance). In temperate regions, on the other hand, less emphasis is placed on parasite resistance and more emphasis is placed on other traits.

Population Structure and Breeding Objective

In the process of determining the best animal, you might ask, "Best for whom?". The answer to this question depends on the function of the animal, the structure of the population and the role of the " breeder" 2 within that structure. Most populations can be thought of as having a pyramidal structure: a relatively small number of breeders at the top selling breeding stock to a larger number of multipliers who in turn sell animals to a great number of end users.

The pyramid suggests a flow of germ plasm – genetic material in the form of live animals, semen, or embryos – from the top down, the elite breeders producing the most advanced animals, breeders at the multiplier level replicating those animals, and end users benefiting from the genetic improvement occurring at the higher levels. Ideally, breeders at each level try to produce animals that will be in the greatest demand by their customers at the next level down, with the ultimate result that the best animal is the animal that is the most useful or profitable for the end user. *End users* can thus be defined as the individuals whose particular needs should form the basis for determining breeding objectives.

In food and fibre producing species (sheep, cattle, swine, and poultry), the end users are commercial producers. These are the persons whose primary products are commodities for public consumption. Commercial dairy farmers produce milk; commercial swine producers produce pork; commercial poultry farmers produce eggs, chicken and turkey. Commercial producers are in most cases not the end of the production chain; beyond them are the processors (dairy plant, slaughterhouses), the retailers and consumers. But the commercial producers are end users because their particular needs reflect the requirements of the entire production chain. They need animals that are physically and reproductively sound, healthy and perform efficiently in their environment.

They also need animals that possess the product and performance characteristics required by the retailers and consumers. The importance of these latter characteristics should be reflected – when the market systems functions well-in the prices paid to the commercial producers for their products. In the Western world, the interest of consumers in the system of production has increased over time. This increased awareness of consumers has resulted in an increased emphasis on health and welfare traits in the breeding objective of farm animals and reduced emphasis on primary production traits (e.g. amount of milk, growth rate and litter size). The breeding industries for recreational and companion animal species (horses, dogs, cats, etc.) differ somewhat in structure from the livestock industries.

The pyramid arrangement is still present, and markets for specialised types of animals exist, but seedstock/commercial divisions are usually less clear and the end users may not be breeders at all. Consider, for example, Labrador retrievers. The end users of Labs are hunters and pet owners. These persons may or may not choose to breed their animals, and the qualities that are important to them are those that contribute to retrieving ability, companionship, health, aesthetics, or some combination of these producers does not really fit here because no consumable commodity like meat, eggs or milk is being produced. The various horse industries provide similar examples. End users of horses range from owners of the most valuable racing animals to causal riders to those that keep miniature horses as pets.

How are Animal Populations Improved?

The purpose of animal breeding is not to genetically improve individual animals-once an individual is conceived, it is too late to change the genotype of that animal-but to improve animal populations, to improve future generations of animals. To this task breeders bring two basic tools: selection and mating. Both involve decision-making. In selection, it is decided which individuals become parents, how many offspring they may produce, and how long they remain in the breeding population. In mating, it is decided which of the males we have selected will be bred to which of the females we have selected.

Selection

Selection is used to make long-term genetic change in animals. It is the process that determines which individuals become parents, how many offspring they may produce, and how long they remain in the breeding population. Most of us are familiar with the term natural selection. Natural selection is the great evolutionary force that fuels genetic change in all living organisms. We commonly think of natural selection as affecting wild animals and plants, but in fact it affects both the wild and domestic species. All animals with lethal genetic defects, for example, are naturally selected against-they never live to become parents. Natural selection cannot be ignored but the kind of selection of primary interest in animal breeding is artificial selection.

The idea behind selection is simply this: to let individuals with the best sets of genes reproduce so that the next generation has, on average, more desirable genes than the current generation of animals. The animals with the best sets of genes are said to have the best breeding values. They are –from a genetic point of view-the individuals with the greatest value as parents. In selection, we try to choose those animals with the best breeding values: the animals that will contribute the best genes to the next generation. The result of successful selection is then to genetically improve future generations of a population by increasing over time the proportion of desirable genes.

To see how selection works, consider the simplest form of selection: phenotypic selection or mass selection. In this type of selection, the performance of the individual is the only information used in making selection decisions. No attention is paid to the pedigree of the animal or the performance of its sibs (brothers and sisters) or of any progeny it may have produced. For example, if you were using phenotypic selection for weaning weight to determine whether a particular ewe lamb was to be kept for breeding, you would base your decision strictly on her own weaning weight. In practise (meaning outside of scientific laboratories), phenotypic selection in its pure form is increasingly rare, but it makes a good example, as we will also see later on during this course.

Multiple Trait Selection

In this course, a lot of the discussion of selection and the examples used for illustration will be limited to single-trait selection, selection for just one trait. That is because single-trait selection provides a simple framework within which to learn the principles of animal breeding. But in the real world of animal breeding, selection for a single trait is rare. Breeders are typically interested in improving a number of traits. They practise multiple-trait selection. Dairy farmers select for traits related to milk production, health, reproduction, type and longevity. Selection for one trait rarely affects just that one trait. Usually other traits are affected as well.

Genetic change in a trait resulting from selection on another trait is termed correlated response to selection. Correlated response to selection is probably caused by a number of genetic mechanisms and results in so-called genetic correlation between traits. Genetic correlations between traits and the correlated response to selection brought about by them can be beneficial. However, if we are unaware of or choose to ignore unfavourable genetic correlations, selection for one trait can lead to undesirable response in others. In cattle, for example, blind selection for growth rate leads to larger birth weights and more dystocia. If we want faster growth, but cannot tolerate increased dystocia, we must avoid simply selecting for growth or against dystocia. We need a way to select for growth rate and against dystocia at the same time. We need a method for multiple-trait selection as introduced in this course.

Inbreeding

Inbreeding is the mating of related individuals. That is the simplest definition anyway. Because all animals within a population are related to some degree, a more technically correct definition of inbreeding is the mating of individuals more closely related than average for the population. Inbreeding has a number of effects, but the chief one and the one from which all the others stem is an increase in homozygosity- an increase in the number of homozygous loci in inbred animals and an increase in the frequency of homozygote genotypes in an inbred population. Because inbred individuals have fewer heterozygous loci than non-inbreds, they cannot produce as many different kinds of gametes. The result is fewer different kinds of zygotes and therefore less variation in the offspring. This illustrates, as we will see in more detail furtheron in this course, that inbreeding (more precisely the level of inbreeding in the population) is related to the amount of genetic variation. A second consequence of inbreeding is the expression of deleterious recessive alleles with major effects, and it is this aspect of inbreeding, more than any other, that gives inbreeding a bad reputation. People associate inbreeding with genetic defects.

It is true that defects caused by recessive alleles often surface in inbred populations. But inbreeding does not create deleterious recessive alleles; they must already have been present in a population. Inbreeding by itself simply increases homozygosity, and it does so without regard to whether the newly formed homozygous combinations contain dominant or recessive alleles. It therefore increases the chance of deleterious alleles becoming homozygous and expressing themselves. Expression of deleterious recessive alleles with major effects, particularly lethal genes, is a very visible consequence of inbreeding. It is an example of the effect of inbreeding can have on certain simply-inherited traits. Less obvious is the expression of unfavourable recessive alleles influencing polygenic traits. The individual effects of these genes are small but, taken together, can significantly decrease performance-a phenomenon known as inbreeding depression.

Biodiversity

An important issue arises in situations where a breed that is native to a particular area appears to have lost its function in that area or elsewhere, and consequently is in danger of becoming extinct. The question to be raised in this situation is whether such a breed should be preserved. The arguments in favour of preservation are that we do not know what type of animals will be required in the future, and that we should therefore preserve the available genetic variation between breeds (bio-diversity) as an insurance against the unknown future. On the other hand, it is argued that people who aim to earn a living from animals cannot afford to look too far into the future; they appreciate the arguments in favour of preservation, but are unable to meet the relatively high cost of preserving populations that they are unlikely ever to utilise during their own lifetimes. At both the national and international level, e.g. FAO and Rare Breeds International, concerted efforts are being made to gather relevant data on breeds that seem threatened by extinction, and to act, where possible, to save them. Interestingly, the two areas that are probably of greatest concern are at the either end of the spectrum of animal improvement. At one end we have a large variety of locally adapted native populations (often in developing countries) that are under threat from the influx of "improved" breeds and strains from developed countries. And at the other end we have an increasing number of poultry selection lines that are discarded when yet another independent poultry breeding company is taken over by a larger and often multinational breeding company.

Technology and Animal Breeding

The face of animal breeding has changed significantly over the past decades. Animal breeding used to be in the hands of a few distinguished " breeders", individuals who seem to have specific arts and skills to " breed good animals". Nowadays, breeding in particular in livestock species is dominated by science and technology. In some livestock species, animal breeding is in the hands of a few large companies, and the role of the individual breeders seems to have decreased. There are several reasons for this change. Firstly, the breeding industry has adopted scientific principles. Looking was replaced by measuring, and an intuition was partly replaced by calculations and scientific prediction. Other major developments grew from the introduction of biotechnology. Biotechnology can be broadly defined as the application of biological knowledge to practical needs. These technologies fall generally into two categories, reproductive and molecular. Not all of this is new. Artificial insemination was introduced in cattle in the fifties. There is no doubt that technology had a major impact on rates of genetic improvement in dairy cattle and is just as important to the structure of animal breeding programs. Nowadays, technologies like ovum pick up, in vitro fertilisation, embryo transfer, cloning of individuals, and selection with the use of DNA-information is all on the ground. Some of the technologies are already applied, others are further developed, or waiting application. Finally, rapid development of computer and information technology has greatly influenced data collection and genetic evaluation procedures in animal populations, now allowing comparison of predicted breeding values across farms, breeds or countries.

It is important to recognise that the introduction and exploitation of new technologies have large social impacts. The introduction of breeding methods typically needs to find the right balance between what is possible from a technological point of view and what is accepted by the decision makers and users within the socio-economic context of the production system. Ultimately it is the consumer who decides which technology is desirable and which is not. In most western societies, consumers are increasingly aware of health, environmental and animal welfare issues. Food safety and methods of food production are part of their buying behaviour. However, price and production efficiency are still major factors determining the sustainability of a livestock sector. Successful animal breeding programs need to find and apply the accepted technologies that help them remain competitive. This course is mostly concerned with the technical issues involved in the application of new technologies in animal breeding.

Components of Breeding Programs

Very generally, the aim of animal breeding is to genetically improve populations of livestock so that they produce more efficiently under the expected future production circumstances. Genetic improvement is achieved by selecting the best individuals of the current generation and by using them as parents of the next generation. A breeding program is the organized structure that is put into place to genetically improve livestock populations. This chapter deals with the set-up and evaluation of animal breeding programs.

5

Applied Animal Breeding for Different Species

Introduction

Pigs are used mainly for producing human foods. Meat cuts are the main interest, but other products derived from the carcass, such as legs and noses (e.g. for Chinese market), are used for human consumption. Secondary uses of pigs include manure production and the fulfilment of cultural needs. In medical research, pigs are also used as models of humans. Pigs are kept in a broad spectrum of production environments around the world, but in Denmark the vast majority are kept in intensive housing conditions with a controlled climate; a minority of Danish pigs are kept outside in free range environments.

Denmark is among the world's largest pig producers. In 2009, 19.3 million pigs were slaughtered in Denmark, which corresponds to 2 million tonnes of meat. Worldwide, about 93 million tonnes of pig meat was generated by slaughter in 2009. In Denmark, 94% of the meat produced in 2009 was exported; and Germany (30% of that meat), United Kingdom (15%), Japan (7%) and China (7%) were among the larger importers of Danish pig meat (Landbrug og Fødevarer, 2010).

Artificial insemination (AI) with fresh (non-frozen) semen is used in most matings. Boars can produce about 50 doses of semen per week, and this allows them to be intensively selected. Purebred Landrace, Yorkshire and Duroc sows farrow 15.3, 15.3 and 9.8 piglets per litter on average. Gilts reach sexual maturity at 6–7 months of age, and their average gestation length is 116 days.

Breeds

Danish pig production is based mainly on three breeds: Duroc, Landrace and Yorkshire. Duroc is used as a terminal sire on Landrace x Yorkshire (LY) sows to produce crossbred pigs for Danish production herds. Other countries use breeds with the same names and similar origin as these 'Danish' breeds, but their populations differ as the result of, among things, different breeding goals and the restricted exchange of genetic material. Hampshire, Piétrain and Berkshire are also used in some countries, and locally other breeds continue to have some commercial influence. China, the world's largest swine industry, has been based on roughly six types of pig, defined by geographical location and origin. However, a rapid transition is taking place in China to US and/or European breeds, and now Piétrain, Duroc, Landrace and Yorkshire are the most commonly used breeds in modern cross-breeding systems. Durocs were imported from North America to Denmark in the late 1970s. Besides its high growth capacity, good carcass traits and high feed efficiency, the breed is recognized by its red-brown colour. Yorkshire and Landrace are both white. They are known for their maternal qualities (i.e. they have large litters and nurse their piglets well).

Breeding Goal

The breeding goal is to breed pigs that will generate the highest possible economic return for commercial pig producers over the coming 5–10 years. This breeding goal is decided by commercial pig producers with the guidance of the Breeding & Genetics section at the Danish Agricultural and Food Council. Economic values for most traits are based on a bioeconomic model. This model simulates incomes and costs of each trait in a 'future' production herd; it can be amended to reflect political concerns. The breeding goal is different for paternal (Duroc) and maternal (Landrace, Yorkshire) breeds.

Genetic Evaluation and Parameters

Multiple-trait animal models are used in the genetic evaluation of groups of 2–4 traits. For instance, estimated breeding values (EBVs) for feed efficiency, the two growth traits and lean meat percentage are calculated using a four-trait model. Although genetic correlations are relatively small this is especially advantageous for feed efficiency, because animals without records on feed efficiency, but with records on one or more of the other traits, obtain EBVs that are based on correlated information.

The explanatory effects used in the genetic evaluations to account for environmental effects differ from trait to trait. Typical effects are sex, herd-year-month of registration, common environment for litters, common environment effect for the housing group of pigs, and weight of the animal at the onset of the registration period (e.g. growth 30–100 kg). The bivariate model for number of piglets alive after day 5 and litter size also includes effects of parity of sow, the sow's age at 1st farrowing (1st parity only), farrowing interval (later parities only) and type of fertilization (AI or natural). The parameters used in genetic evaluation and in the breeding programme for Landrace pigs are summarize. The heritabilities and correlations are similar for Duroc and Yorkshire, whereas variances in some traits differ. Strength of legs and claws, number of pigs alive after day 5, and sow longevity have low heritability (0.08–0.17). The last two traits are not evaluated for Duroc. Genetic correlations between male (e.g. growth, feed efficiency) and female traits (e.g. no. piglets alive after day 5, sow longevity) tracked in the Danish system are not estimated. Research on foreign pig populations suggests that the genetic correlations between growth and reproductive traits are either unfavourable (e.g. Holm et al., 2004) or close to zero (e.g. Arango et al., 2005).

Organization and Breeding Programme

Danish pig breeding is organized around a classical breeding pyramid. In 2010 the Danish pig population consisted of 32 breeding herds (1785, 2210 and 2717 Duroc, Yorkshire and Landrace sows, respectively), 153 multiplier herds (69 700 purebred sows) and 2601 production herds with 1.1 million crossbred sows. There is some overlap between the figures, as 29 breeding herds are also multiplier herds. The breeding herds form a closed nucleus with no imports from lower tiers in the pyramid or foreign populations. Thus it is only selection and mating decisions made in the breeding herds that influence the additive genetic trends in the population. The current average genetic level in production herds corresponds approximately to the average genetic level observed in the breeding herds 1–2 generations ago. (Transmission of genes takes 1 and 2–3 generations for boars and sows, respectively.)

Breeders send their best boars to AI-stations and also sell approximately 1000 (mainly Duroc) boars per year to production herds. Purebred females are sold to multiplier herds and, in some cases, directly to production herds. Hence, breeders successfully breeding superior pigs earn more money than their less successful competitors. This is an important motivation for breeders to do their best when they record breeding goal traits, selecting animals with the best EBVs and ensuring optimal matings. The main function of multiplier herds is to facilitate the transmission of genetic progress made in breeding herds to production herds. In practice, this means producing crossbred females (LY) that are sold to production herds. Multiplier herds receive purebred Landrace and Yorkshire females from breeding herds. Breeding decisions in production herds are not relevant to future generations of the pig population. Such herds exist primarily for the production of pigs for slaughter. As the vast majority of pigs are raised in production herds, the breeding goal should reflect the circumstances in production herds, and ideally performance measures of breeding animals should be carried out in similar production environments.

Most traits are recorded in the purebred breeding herds. However, feed efficiency is recorded at the test station 'Bøgildgaard' and not in individual herds. The number of piglets alive per litter is recorded in multiplier herds as well as in breeding herds to provide sufficient accuracy of breeding values (trait only expressed by sows and low heritability). Slaughter loss is only recorded for slaughtered animals, which makes it impossible to have own records on active breeding animals. The remaining traits are recorded on most pigs in breeding herds — only approximately 25% of the pigs do not have their performance recorded, and this is due mainly to death, disease or experimental discrepancies. The approximate proportion of tested pigs that are used for pure-breeding. Selection intensities are substantially higher for boars than gilts as a consequence of AI being used. These intensities are lower for Duroc as compared with the maternal breeds as a result of Duroc's smaller average litter size, smaller population, and because some Duroc boars are used for both breeding and production herds. The use of selected boars varies substantially (i.e. the number of matings per boar ranges from 1–60).

Inbreeding only concerns breeding herds (i.e. purebred pigs) and is controlled by imposing an upper limit of 50–60 matings for a single boar, depending on the breed. Furthermore, a maximum of 40 half- and 2 full-brothers are accepted at the 'Bøgildgaard' test station for Yorkshire and Landrace, whereas a maximum of 100 half-and 3 full-brothers are accepted for Durocs. Breeders decide which matings to arrange on the basis of these limitations. Limiting the use of each boar is easy enough in practice, but it is not an optimal way of controlling inbreeding since it does not account for relationships among boars and their breeding values. Therefore, The Danish Agricultural & Food Council's Pig Research Centre is working on implementing optimum contribution selection of boars (Bendtsen, 2008).

Genomic EBVs based on a 62K SNP chip are currently being developed for all evaluated traits. They are expected to have the greatest impact on longevity, litter size and feed efficiency, where accurate EBVs are not available for young selection candidates. Conversely, they are expected to have little impact on the remaining evaluated traits. Potentially, genomic EBVs will also be developed for health traits that are not being evaluated today. Furthermore, genomic EBVs permit the collection of data on crossbred sows — and the subsequent use of this information, in connection with purebred animals. This helps to overcome problems with genotype by environment interactions and gene expression differences between pure and crossbred pigs due to different background genetics.

Examples of Genetic Trends

Favourable genetic trends for growth, feed efficiency and number of piglets alive at day 5. Duroc has made greater genetic improvements in growth and feed efficiency than Landrace and Yorkshire. This can be explained by the higher relative emphasis on these traits in the breeding goal and the larger number of animals with records for feed efficiency. Genetic progress has been lower in Yorkshire than in Landrace for these three key traits; this could be due to the former's smaller population size, small differences in genetic variance or chance. On the other hand, Yorkshire has improved more than Landrace in lean meat percentage over the same period (results not shown).

Dairy Cattle Breeding

Milk production is the main purpose of dairy cattle production. The ideal cow will have a high milk yield; the milk should have a certain quality (e.g. protein and fat content, low somatic cell-count) and the costs of milk production should be low. A secondary benefit of dairy cattle farming is beef production, although this is not valued in the breeding goals of all breeds. Cattle may also be used for nature conservation, their skin for leather, their bones for various tools, and in some cultures cattle are important for cultural or religious (e.g. Hindu) reasons. Cattle are able to convert inexpensive roughage that cannot be used directly as human food into human food. However, dairy cattle feed in western countries also contains high concentrations of grain, and this is of concern to some people. Other public concerns about cattle include their welfare and carbon-gas emissions.

Dairy cattle are kept in a broad range of environments from low-input pasture to high-input tie-stalls or free-stalls. The typical Danish dairy herd a couple of decades ago was tie-stalled and numbered less than 50 cows. Today most cows in Denmark are housed in free-stalls in herds of more than 100 cows. At the same time automatic milking systems are replacing more labour intensive parlours in many north-western European countries where there are high minimum wages. Large herds, where animals receive homogeneous management within the herd, make it easier to correct for non-genetic effects in genetic evaluations; they are therefore beneficial for efficient breeding.

About 1.3 billion (i.e. $\times 10^9$) cattle are spread around the globe, if we include both dairy and beef, and this number has been fairly constant for the past few decades (FAO, 2008). In Denmark and other European countries the number of dairy cows has decreased over the past few decades, but the amount of milk being produced has been fairly constant due to a steady increase in milk yield per cow.

The reproductive rate of bulls is high by nature, but can be made extremely high through the cryopreservation of semen and artificial insemination (AI). Hence, a few elite bulls are used in several countries and these have sired several hundred thousand offspring worldwide. By contrast, the reproductive rate of females is low. A cow can have its first calf at around 2 years of age and will on average have just one calf a year hereafter. However, heifers and cows can be super-ovulated using hormones to produce more embryos (around 5 to 6 transferable embryos on average per flush, but variability is high). These embryos can subsequently be flushed and transferred to recipient cows that carry the pregnancy to term. This technique is often used with elite cows to increase their reproductive capacity. It is also possible to separate 'male' and 'female' sperm cells so that the sex of the offspring can be chosen with an accuracy of over 90%.

Breeds

However, in western countries with developed dairy production there are fewer than 10 principal international breeds. Here the Holstein population is by far the largest, followed by Simmental, Red Dairy Cattle (RDC), Jersey and Brown Swiss. A great deal of genetic material is exchanged among countries (primarily via semen), so distinguishing between 'national breeds' makes little sense. For instance, many different North American Holstein bulls have been used for decades, and the proportion of genes that can now be traced to original Danish Black and White Cattle (SDM) was estimated at 1.6% in Danish Holstein calves born in 2009. This upgrading process has been termed 'Holsteinization'; it has occurred in many countries.

Holsteins yield large quantities of milk in high-input production environments; they have udders that are well suited to modern milking systems. However, they are considered less robust in extensive or stressful production environments. They are present in all of the main dairy countries. Simmental was originally a dual purpose breed, but today separate lines focus on either beef or milk production. Milking Simmentals are very popular in central Europe and especially in mountainous regions. Red Dairy Cattle are popular in Nordic countries. A relatively strong emphasis has been placed on functional traits such as health and fertility in their breeding goal. Jerseys are a smaller breed and Jersey milk carries a relatively high concentration of fat and protein. The largest Jersey populations are found in New Zealand, the USA and Denmark. In Denmark there are 590 thousand dairy cows (Danmarks Statistik, 2010); most are Holsteins (72%), followed by Jerseys (12%) and Red Dairy Cattle (RDC; 8%).

Breeding Goal

The three Nordic countries, Denmark, Finland and Sweden, have joint breeding programmes and the same breeding goals, but the breeding goals differ slightly across the breeds. Breeding goals are decided upon by each breed association. However, to a large extent the associations base their decisions on results from analyses using bio-economic models of revenues and costs in a typical future herd. The Nordic Total Merit (NTM) index reflects the breeding goal of the particular breed. The traits mentione are in most cases sub-indexes made up of several individual traits, as described in the footnotes to the table. The value of a one-unit increase in the NTM index corresponds to 75, 58 and 67 DKK per cow per year for Holstein, Red Dairy Cattle and Jersey, respectively.

Feed efficiency is not assigned a direct economic value in the NTM indexes, because no registrations are available for feed efficiency. However, feed prices are taken into account in the bio-economic model, and they influence the economic values for dairy production. Hence, when feed relative to milk prices increases the economic values for functional traits relative to dairy production also increase. Such modelling is based on many critical assumptions. It could never be as sound as the kind of model that would be based on the inclusion of direct economic values if feed efficiency records were available on individual animals.

The breeding goals in the Nordic countries place a relatively strong emphasis on functional traits, and especially on health, due to high labour and veterinary costs. Non-Nordic countries keep no systematic records of health traits.

Genetic Evaluation and Parameters

Nordic Cattle Genetic Evaluation (Nordisk Avlsværdivurdering, NAV) computes EBVs jointly for Denmark, Finland and Sweden four times annually. Multiple-trait models are used for the traits belonging to the sub-indexes mention. For instance, milk, fat and protein yield from 1st, 2nd and 3rd parity are treated as different traits using the estimated correlation structure among traits. Likewise mastitis in different parities and lactation stages is treated as a plurality of traits and analysed simultaneously with somatic cell-count, fore udder attachment and udder depth to increase the accuracy of predictions. Longevity, conformation and workability traits (milk ability and temperament) are analysed in single-trait models. Animal Models are used for dairy production, beef production, udder health, longevity, conformation and workability traits. Sire Models are used for the remaining traits. This means that EBVs for important functional traits are not available for females, except when they are based on paternal pedigree information. In the near future, genetic evaluations for claw health based on records from claw trimmers are expected.

The statistical models used for genetic evaluations include a number of different explanatory effects, such as herd, year, month and age associated with the given performance. These effects may be different for different traits depending on the frequency of measures, heritability, the selection emphasis, the biology of the traits and the evaluation method. For instance, daily yields are measured on different test-days, so the model for dairy production includes the effect of specific test-days rather than a monthly measure. The model for dairy production also accounts for the shape of the lactation curves, which makes it complicated. Further details of the evaluation models for specific traits can be found in the Danish Knowledge Centre for Agriculture (2010; in Danish) or Interbull (2010; for several countries).

Genetic parameters for different traits (or indexes) in the breeding programme are summarize for Holsteins. Heritabilities and correlations are similar in other breeds, whereas variances differ more. Generally dairy production has unfavourable genetic correlations with functional traits. The genetic correlations were approximated from EBV. A few of the estimates were more extreme (e.g. for dairy production with female fertility and other diseases, respectively) and others closer to zero (e.g. for longevity with female fertility and other diseases, respectively) compared with similar estimates in the literature.

International genetic evaluations for bulls are conducted three times annually by the organization Interbull so that objective comparisons of bulls across country borders can be made. All traits mentione are evaluated, except beef production. A multiple-trait model is used within which performance in each country is considered a distinct trait, thereby allowing for country-specific selection according to own production circumstances. Across-country genetic correlations between milk yield in similar production environments such as Denmark and the Netherlands (0.92) are stronger than they are in less similar production environments such as Denmark and New Zealand (0.75), because cows in New Zealand are on pasture all year unlike those in Denmark and the Netherlands.

Genomic breeding values for genotyped animals have been calculated for all of the traits in the Nordic breeding goal since 2009. Other leading dairy countries have also implemented genomic predictions or are in the process of doing so. So far these calculations have been based on the traditional EBV, or functions of it, but a method integrating all of the information in one step is being developed. Practical results indicate that the reliability (r^2_{IA}) of genomic EBV ranges between 30–72% for all traits considered in the Nordic countries and between 41–53% for welfare traits; these figures are substantially higher than the average for parents' breeding values (Su *et al* 2009).

Organization and Breeding Programme

VikingGenetics is a farmer owned organization that is responsible for practical cattle breeding in Denmark, Finland and Sweden. Its responsibilities include selecting and testing bulls and producing and marketing semen, as well as conducting AI and advising farmers on breeding (e.g. on insemination plans).

Until recently dairy cattle breeding has not followed the breeding pyramid structure we see in pig and poultry breeding. This is mainly due to the low reproductive rate of females. Instead nearly all cows in the population are recorded and potentially available for breeding.

Nor is crossbreeding widely used in Denmark (less than 10% of Danish dairy cows are crossbred), but it is widely used in New Zealand and is increasing in the USA. Widespread use of AI characterizes dairy cattle breeding, but the main traits are only expressed in females where the selection intensity is low because most of the heifer calves born are needed to maintain a constant population size. This has resulted in long generation intervals for sires, because intense bull selection was not optimal until progeny information was available. (Bulls were at least 5 years old when first daughters went into first lactation.) The breeding system used for decades in many countries, including Denmark, was a 4-pathway structure concerned with selection of sires for sires (SS), dams for sires (DS), sires for dams (SD) and dams for dams (DD).

In Danish Holstein, approximately 5–6 SS and 10–12 SD were selected per year from the 240 progeny-tested bulls. Similar proportions were also selected from 60 Swedish and 50 Finnish progeny-tested bulls. The selected bulls were 5–7 years old. At the same time about 2000 DS and 90% DD were selected from the cow population. The selected DS and DD are younger than SD and SS when they are selected, but before genomic EBV became available they had typically had at least one own lactation (i.e. were >2–3 years old). More SS than are needed are selected to avoid accelerated rates of inbreeding. Also, individual farmers avoid mating close relatives. To facilitate this VikingGenetics looks for bulls with alternative pedigrees in addition to EBVs when selecting SS and young bulls for testing.

Today the 4-pathway breeding system is being revised in responses to the recent availability of accurate genomic EBV at an early age. Countries have been quick to adopt this new technology, but little is still known about how it is best used to enhance genetic progress with low risk. Initially, then, more or less conservative modifications of the traditional 4-pathway structure are being implemented. In Nordic countries genomic EBV is presently used to intensify the pre-selection of young bulls to be progeny-tested and to identify bull dams (DS) for super-ovulation and embryo transfer. Thus, heifers with high genomic EBV are used as DS; young bulls with high genomic EBV are also being used to some extent. Although fewer young bulls may start progeny testing than before, the screening is expected to be much more accurate than the previous screening, which was based primarily on the average EBV of parents. The following changes, which were made in the Nordic Holstein breeding plan after genomic EBV became available, illustrate recent developments (Lars Nielsen, VikingGenetics, pers. comm.):

Young bulls (1.5–2.5 years old) with high genomic EBV (GenVikPlus) were used for 15% of all inseminations in August 2010About 1300 young bulls are genotyped and 225 of these initiate progeny testing. Hence, about 33% fewer young bulls are progeny-tested than previously. In the future even fewer Holstein bulls are expected to be progeny-tested due to closer cooperation with other European Holstein populations (EuroGenomics). Other breeds do not have the same opportunities for cooperation, and therefore with these breeds it is not possible to reduce the number of progeny-tested bulls as much in order to maintain a sufficiently high accuracy of genomic EBV

Up to 10 000 semen doses of each of the most promising young bulls, based on genomic EBV, are used immediately and approximately 10 000 semen doses are savedVikingGenetics genotype about 500 Holstein heifers and cows. In addition to this private farmers genotype some females. About 20% of the waiting bulls (with initiated progeny testing, but awaiting progeny results) with the lowest genomic EBV are culled

In the future the availability of genomic EBV may lead to more fundamental changes in the breeding programme. For instance, although high quality data on a reasonable number of animals (the precise number depends on heritability, population structure, and so on: see the notes on genomic selection) are always crucial, data records for *all* animals are no longer required. Instead specific herds may be targeted for more intense data recording. Through the extended use of embryo transfer and sexed semen an open nucleus scheme with systematic crossbreeding in production herds can be envisaged.

Examples of Genetic Trends

In the 1990s most of the selection emphasis was on dairy production — at least, in reality, due to the heavy use of North American bulls. At the beginning of the new millennium the emphasis on functional traits such as udder health, longevity and female fertility increased. The onset of international genetic evaluations for these traits in 2001, 2004 and 2007, respectively, made it easier for the participating countries to select foreign bulls according to their breeding goals, especially for the Nordic countries which puts relatively strong emphasis on functional traits compared with other countries. By comparison, genetic trends for functional traits were not unfavourable for Red Dairy Breeds as they relied to a large extent on Nordic bulls.

Poultry Breeding

Industrial poultry breeding involves a wide range of species (chickens, turkeys, ducks, geese, quail, and ostriches) and purposes (eggs, meat, feathers, leather, and oil). Both in Denmark and worldwide it is chickens — for eggs (layers) and meat (broilers) — that are most numerous and economically important bird in industrial poultry breeding.

Worldwide, turkey and duck breeding also operate on quite a large scale and are important, whereas geese, quail, and ostrich breeding can generally be categorized as small-scale industrial breeding for niche production. The production size and number of herds in Denmark and the worldwide production size are given for various industrial poultry breeding species.

The reproductive capacity of poultry is generally high. This is especially so with chickens, which have a particularly high female reproductive capacity compared with other species. This gives poultry some of the shortest generation intervals in farm animal breeding. Another important feature of poultry breeding is that males are homogametic (ZZ), whereas the females are heterogametic (ZW). This affects which reproductive technologies are possible and also crossbreeding organization.

Breeds

For a number of species the most important breeds used in industrial poultry breeding. It is worth noting, however, that in most major poultry breeding operations the term *breed* is now rarely used. Rather, the terms (pure or pedigree) *line* and (final) *product* are used.

Lines originate from one or more breeds; they are bred and selected in closed populations. A distinction is made between so-called sire and dam lines — which, in species kept for meat production, usually originate from different breeds.

Products are the birds that are used in the final market operation. In layers, for instance, they are the birds that produce eggs for consumption, and in species kept for meat production they are the birds that are produced for meat consumption. Products are usually line crosses, in some cases between lines originating from different breeds.

Breeding Goal

A detailed listing of traits included in the breeding goals of various industrial poultry breeding species. Comprehensive studies of the economic and social values in poultry breeding programmes have not been published, except for economic values in broilers. Examples for some traits are provide, but values will depend on the market (e.g. battery or floor-housing egg production, and live bird or processed meat production). Many poultry breeding companies are applying a desired gains approach rather than using economic values based on cumulative discounted expressions. The general breeding goal for laying hens includes traits related to: high number of saleable eggs per hen, feed conversion, egg quality, mortality, and adaptability to specific commercial environments. Interest in adaptability increasingly focuses on floor management and involves traits such as nesting behaviour, feather pecking, and cannibalism.

For broilers the main breeding goal is, and has always been, body weight at slaughter age. Many other traits are also included in the breeding goal, though. For example, feed conversion, slaughter yield, mortality, leg health and cardiorespiratory health, and female and male reproduction traits. In recent years, quality traits have also been included for some markets. These include one or several of the following: body conformation, intramuscular fat, tenderness, drip loss, plumage-, skin-, and shank colour, comb redness and size, and feathering rate. The breeding goals for turkey, duck and geese are generally similar to those for broilers. They exclude cardiorespiratory health, however, as this problem chiefly affects the very intensively selected broilers.

The main driver in duck breeding has been consumer demand for lower-cost food products. The principal foci of the breeding goal were initially (1970s to early 1980s) body weight, laying performance and survival ability; but afterwards (until 2000) the focus became broader, as changes in housing systems and breeding industry expansion occurred. The breeding goal now came to include traits such as feed conversion ratio, body composition (high meat and low fat yield), leg strength, fertility, hatchability, egg weight and eggshell quality. Since 2000 even more traits have been included in the breeding goal, including sexual behaviour traits (fertility) and mobility (leg strength), but disease resistance and health disorders have only very recently (2006 on) come under consideration. Specialized breeding companies focusing on foie gras production have a slightly different focus (not low fat yield, but high liver yield). If economic values are used, none have been published. It is possible that a desired gains approach is used rather than economic values. In ostrich breeding, the economically most important traits serve leather and meat production, and to some extent feather quality. In recent years, there has been a shift towards meat production. The relative importance (b-values) of leather, meat, and feathers was approximately 70%, 25%, and 5%, respectively, about a decade ago; today the corresponding figures are 45%, 45%, and 10%.

The distinction between the sire and dam lines mentioned earlier is reflected in breeding goals. The sire line is a line bred with the purpose of obtaining an outcome in the male at the parent stock level in the breeding pyramid, whereas the dam line is bred with the purpose of obtaining an outcome in the female at parent stock level. This differentiation in of the breeding goals involved in sire and dam lines should ensure that there is positive heterosis in the final product. In the breeding goal of layer sire lines relatively large weight is placed on egg quality and male reproduction traits. In the breeding goal of broiler sire lines, by contrast, more weight is attached to production traits such as body weight and breast meat yield, as well as to male reproduction traits. In both layer and broiler sire lines relatively little weight is placed on female reproduction traits. In the breeding goal of layer dam lines the main focus is on female reproduction traits. In the breeding goal of broiler dam lines relatively little weight is placed on production traits such as body weight and breast meat yield and more weight is placed on female reproduction traits.

Genetic Evaluation and Parameters

In the larger poultry breeding companies targeting conventional markets, genetic evaluations are mainly conducted using BLUP animal models, although phenotypic culling is also applied sometimes, especially for functional traits. In some companies, multi-trait models are used to the extent allowed by computer processing limitations, but some companies still apply single-trait models. Advanced models, including non-additive genetic effects (e.g. common environment, maternal or heterosis effects) are used by some companies, and the inclusion of genetic markers is gradually being implemented by various companies. Smaller (and often local) breeding companies apply mostly mass selection based solely on phenotypic information.

Examples of typical genetic parameters for important traits in layers and broilers, respectively. The number of studies reporting genetic parameters for important traits in turkey, duck, geese, and ostrich breeding is limited, but the studies published so far report heritabilities that are generally in the same range as those reported for broilers — except for heritabilities of body weight in geese, which were very high.

For genetic evaluations in the pure lines of large companies (breeding layers, broilers, turkeys and, to some extent, ducks) single-trait and/or multi-trait BLUP are used, but selections based on BLUP-EBV are usually accompanied by more or less elaborate phenotypic culling. Great grandparent stock may also be selected on the basis of genetic evaluations, but it is only in the grandparent and parent stock that phenotypic culling is applied. The inclusion, by poultry breeding companies, of genomic information in genetic evaluations is becoming more and more common. It enables selection to be based to some extent on genome-wide breeding values. More companies are expected to join this trend. Molecular genetic information has already been used successfully for simple inherited traits in some poultry breeding programmes. For example, genetic tests are used to identify the presence or absence of specific colour genes in connection with the development of coloured broiler lines for alternative production systems. In layers, an example of a successful genetic test is the test for fishy taint in eggs (the *FMO3* gene). This test has allowed the taint problem to be eliminated in most, if not all, commercial lines today. In smaller scale breeding companies (especially those working with geese, quail and ostriches) classical selection index or even pure mass selection is still practised.

Organization and Breeding Programme

Today layer, broiler, turkey and duck markets deal mainly in the final products of just a few large-scale, centralized breeding companies. For decades the companies have been specializing in, and merging within, each species, and therefore growing. During the past few years, however, there has been a trend for specialized companies to merge into multi-species companies, with a range of products for each species. The major companies include the *Erich Wesjohann Gruppe*, which concentrates on products in the white and brown layer (*Lohmann Tierzucht, Hyline, H&N*) broiler (*Aviagen*) and turkey (*Aviagen, British United Turkeys*) markets; *Hendrix Genetics*, which concentrates on products in the white and brown layer (*ISA, Hendrix*) and turkey (*Hybrid*) markets; the *Grimaud Group*, which concentrates on products in the broiler (*Hubbard*) and duck (*Grimaud*) markets; and *Tyson* (broilers: *Cobb-Vantress*) and *Bangkok Ranch Group* (ducks: *Bangkok*

Ranch, Cherry Valley). The major driver of the formation of these multi-species companies has been collaboration in research projects, particularly in genomics. In Asia and Africa, especially, large parts of the poultry markets are, however, still concerned with local breeds that have been bred in small companies or small-scale holdings — for example, the Yellow Bird, which is a local Chinese meat-bird product.

The typical breeding pyramid structure of the poultry breeding industry. The pure-line elite stock is located at one main location, preferably minimized to one satellite farm. A satellite farm is essentially a backup breeding programme, which is a copy of the central one, but which is located elsewhere to spread any risks presented by, for example, A-list diseases. At the pure-line level, a breeding company typically has a number of commercial as well as experimental lines. The commercial lines are used in final products currently being marketed; experimental lines are developed for potential new products of the future, or for the exchange of lines in current products. A final product is typically a three-way or four-way cross, as shown for broilers. The breeding programme is at the pure-line elite stock level, possibly with simultaneous measurements at great grandparent stock level (which can be used as additional information in the genetic evaluations). The great grandparent, as well as the grandparent, stock levels are, however, primarily multiplier levels. The parent stock level, like the final product level itself, is considered a production level. There are four generations (4–5 years) between the pure-line and final product level. Thus there is always a genetic gap between pure lines and production birds at the final product level.

Currently the potential of lines as either sire or dam lines is limited by the feather sexing procedure used to distinguish 1-day old male and female chicks at final product level (this distinction allows the separation of males and females for a quicker turnover). Feather sexing makes use of a genetically determined differentiation in feather growth. The dominant sex-linked gene, K, results in slow feathering; the recessive allele, k+, results in fast feathering. In slow-feathering chicks the primary wing feathers are short and no longer than the coverts. By contrast, primary wing feathers are longer than the coverts in fast-feathering chicks. In the final product males must be slow-feathering and females fast-feathering, and to achieve this, the sire lines must be fast-feathering and the dam lines used as a male at grandparent stock level (C) must be slow-feathering; dam lines used as a female at grandparent stock level can be either fast or slow.

A multi-stage selection strategy, with at least two selection steps, and an overlapping generation structure is usually adopted in poultry breeding. Typical breeding programmes in layers and broilers are illustrate. To a great extent, these are representative of other meat production species as well.

In layers performance tests of pure lines and crosses (by reciprocal recurrent selection) for feed efficiency (production/feed intake) and reproduction traits, among other traits, are run roughly between 20–50 weeks of age. Reciprocal recurrent (i.e. repeated in each generation) selection is based on the performance of cross-line relatives by assigning sires of each line to be mated to dams of each line and the other way around. The crossbred offspring can then also be performance-tested. There is one pre-selection step (1st step: during rearing) and one or two selection steps during the production period (e.g. 2nd step: after peak production and 3rd step: well through the production period — between 50 and 60 weeks of age — in order to include information on persistency and information from the reciprocal recurrent testing). Matings may be reshuffled after the 3rd selection step. Selection intensities are high, and an approximation of the selected proportions within a generation is that ~0.2–0.5% of males and ~1–3% of females are kept for reproduction of the next generation.

In broilers performance tests are conducted between approximately 3 and 8 weeks of age. They include feed efficiency testing in individual cages, slaughter tests for, among other things, meat yield, and health and performance testing in challenging environments. Testing focuses mainly on the pure lines and only to a limited extent on crossbred offspring. There are one or two selection steps at an early age (in broilers: 1st: 3–5 weeks of age; 2nd: 6–8 weeks of age), where the first (if present) corresponds to a transfer of (a part of) the birds to specialized test facilities, and the second corresponds to the age at which that species would normally be slaughtered as a final product. In addition, there are one or two selection steps for older birds: one prior to the egg-laying period (3rd step) and one (some time) after peak egg production (4th step), where, after matings, birds may be reshuffled. The 1st and 4th selection step are often differentially applied to males and females and are sometimes applied only to males. Selection intensities are higher in the first two selection steps than they are in the last two selection steps. An approximation of the selected proportions within a generation is that ~0.5–1% of males and ~2–4% of females are kept for reproduction of the next generation, depending on, among other things, the reproductive ability of the line (a higher selection intensity is possible in dam lines).

The only reproduction technology employed is AI, but it is widely used. Poultry semen cannot be stored for long, as freezing techniques are not well-developed; nor can it be diluted much. Generally, the number of females per male is therefore no higher when AI is used than it is in natural mating. The main advantage of AI relates to the control, and knowledge, of the full pedigree, and the fact that there is often a higher fertilization percentage in AI than there is in natural mating. The sexing of eggs (or embryos) before, or during, incubation is not currently feasible. A new method of sexing young embryos, which involves determining the dosage of the Z-linked gene DMRT1 in young embryos, is currently being developed for industrial purposes.

Inbreeding is a high risk in poultry breeding, given the high selection intensities applied, and in layers slight inbreeding depressions in sexual maturity and fertility have been reported. Until recently, the avoidance of full-and half-sib matings and the selection of a maximum number of offspring per sire has been the typical strategy deployed to manage inbreeding. In recent years, however, optimal genetic contributions theory is being implemented by larger companies. The poultry breeding industry is perfectly organized for this, as all matings are usually within company control (at least to the extent that AI is used).

Examples of Genetic Trends

In layers a genetic increase in the number of eggs produced of about 1.8 eggs per year was achieved in the period 1950–1993. In the same period, egg mass improved by ~43%, egg weight by ~12%, and feed efficiency by ~32%.

Historically, broilers have been one of the best examples of just how effective traditional quantitative breeding methods can be in obtaining genetic gain. Growth rate in them saw a fivefold increase over the period 1950–2000, with genetic gains in body weight of 58 g per year from 1957 to 1976, 73 g per year from 1976 to 1991, and 84 g per year from 1991 to 2001. In 2010 continued yearly genetic gain of ~50 g body weight at ~6 weeks of age is not unusual (actual genetic gain depends on how much weight is put on other traits in the breeding goal). In the period 1950–1993 carcass yield was improved by 91% and feed conversion by 63%. The genetic gains in some broiler, duck and goose traits are shown. In contrast with the gains observed for layers, broilers and ducks, no genetic gains have been observed in geese; this is probably due in part to the use of sub-optimal methods for genetic evaluations.

The genetic gains in desired traits are accompanied by correlated developments in other traits that are sometimes undesired. For example, in broilers, turkeys and, to some extent, ducks many health-related traits have been negatively affected by commercial selection pressure. These traits relate to the circulatory system (sudden death syndrome and ascites) and the musculoskeletal system (tibial dyschondroplasia, femur head necrosis, deep pectoral myopathy). In layers the worsening of traits such as flightiness, cannibalism and feather pecking, anorexia, and unenthusiastic nesting behaviour are also believed to be connected with commercial selection. Most poultry breeding programmes are now trying to minimize such consequences — for example, by recognizing the problematic traits in breeding goals and setting up tests to gather phenotypic information on them.

Fur Animal Breeding

The main purpose of fur animal production is the production of fur for the textile and fashion industry. Most fur (85%) comes from farmed fur animals, but fur from wild animals (e.g. hunted by indigenous people) is also traded. The most commonly farmed fur animal in Denmark and worldwide is the mink (*Neovison vison*), followed by the fox (including the blue *Alopex lagopus* and the red sort *Vulpes vulpes*). Other species, farmed on a smaller scale, include nutria (e.g. *Myocastor coypus*), chinchilla (*Chinchilla lanigera*), fitch (*Mustela putorius* and *Mustela eversmanni*), sable (*Martes zibellina*) and Finn raccoon (*Nyctereutes procyonoides*). Most fur (both farmed and wild) is sold via international auction houses.

Denmark is the world's largest producer of mink fur. This is mainly due to convenient feed supply from animal by-products (e.g. from the fishing and slaughter industry), the availability of cheap straw, infrastructure required by, for example, feed production, a favourable climate, and tradition. In Denmark the number of breeding mink females has been broadly constant over the past 25 years (and was at 2.7 million in 2009). On the other hand, the number of mink farms decreased from more than 5000 in the late 1980s to about 1400 in 2009 (Clausen, 2010), so the average herd size has grown (and was just below 2000 breeding females in 2009). The latter — global sales — follow the average price per pelt quite well. Other fur producing countries include China, Russia, Ukraine, Canada, the USA, the Netherlands, Finland, Norway, Sweden, South Korea, Poland and Argentina. Major exporting countries such as Denmark have a fur auction (e.g. Kopenhagen Fur). The vast majority of fur sold at Kopenhagen Fur is produced in Denmark, but pelts from other countries such as Sweden are also sold there. Likewise, a few (<5%) Danish produced pelts are sold abroad. China has a large home market and therefore exports relatively few pelts. It therefore does not figure in the data on world production.

Mink are housed in cages holding 1–4 animals. Typically, one male and one female pup are put in a cage together to avoid fights. After pelting, when there is more space available, there is only one female per cage. The minimum size of the cage is regulated by legislation to ensure a certain standard of animal welfare. In Denmark, farmed fur animals must also have access to straw and either a shelf or cylinder. Despite of these environmental enrichments, ethical concerns remain an issue among animal rights organizations over suppressed natural behaviour and injuries from the bites of cage-mates. Illustrates typical Danish mink environments. The main fur animals are carnivores, and their feed contains animal bi-products. Rodent species such as chinchilla, beaver and rabbit are used very little. Mink production follows a fixed seasonal cycle that depends on the reproductive cycle and fur development of the animals (stimulated by changes in the amount of daylight). In Denmark minks are naturally mated in March. The female is in heat for about 3 weeks and each male can only mate with 5–6 females per season. Mating induces ovulation, and the aim is to mate females twice per season. This results in a period of gestation that varies in length from 42 to 72 days, but is generally longer for early matings. Accordingly, most kits are born within 2 weeks around the 1st of May. Mean litter size is approximately 5.5 weaned kits per female (weaning occurs at ~8 weeks).

Breeds

Breeds are usually defined according to fur colour, and indeed the term 'colour type' is often used instead of 'breed'. However, within each colour type there can be several strains with different characteristics. The wild mink is brown, but animals with colours ranging from white to black are a result of mutations in colour genes. The black type, however, is special, as the degree of darkness is a polygenetic trait and the black colour type is therefore a result of several generations of selection. With the exception of black, fur colour type is a qualitative trait based on mutations. At least 25 different loci affect fur colour, many of them with a number of different alleles (mutations). Although the fur colour type is a qualitative trait, there is polygenetic variation within colour type with respect to darkness and colour clarity. In the Nordic countries the development of mink mutations was at its height in the 1960s and 1970s — for foxes it was the 1980s (Lohi, 1993). The production of colour types varies somewhat from year to year, according to fashion trends. When prices for a specific colour type increase the breeders tend to react by preparing more animals with this colour type. Average litter size differs between colour types (Østergaard, 2010), ranging from 5.9 (brown) to 4.5 (violet). Of the more frequent colour types, black is known to have relatively small litter size (5.0). This could be due to the fact that inbreeding is more serious for the black type because it cannot be outbreed with other colour types without hampering the black colour that is a result of several generations of pure breeding. Another possible explanation refers to the pleiotropic effects of alleles for dark colour. The other colour types can more easily be used mutually as the original colour type is easily restored after one generation of backcrossing. Mahogany is a 'synthetic breed' originally created by crossing black and brown animals.

Breeding Goals

Breeding goals are farm-specific. Each breeder decides which traits should be included in selection decisions and what weight should be put on each trait. In practice many breeders use software based on a desired gain approach. They look at the expected realized genetic gains that will be achieved for each trait given the relative weighting factors and the estimated breeding values of animals at the farm. Given the fur colour type, typical traits included in selection decisions are overall fur quality (e.g. hair density, fur purity, hair length, hair elasticity, colour shade and colour darkness), pelt size (selected through body weight, as this and pelt size are strongly correlated) and reproduction (e.g. litter size). Furthermore, several traits such as health, and temperament (e.g. degree of stereotypic behaviour, pelt gnaw) are considered for pre-selection based on a subjective overall phenotypic assessment (which is used because most farmers do not systematically record these traits). All traits in the farm-specific breeding goals are recorded by the farmer and most of them are subjectively scored. For instance, fur quality is scored in

November at the same time as weighting and before pelting. Not all animals have their fur quality scored, as this would be too time-consuming. Feed efficiency is currently not considered and this may cause the value of pelt size to be overestimated. There is some interest in breeding animals that utilize their food better, especially on farms that already register the amount of feed given to each animal. However, this would require an additional weighing of the animals after they have been weaned (e.g. in August). The weight gained relative to the amount of feed provided is a result of both feed utilization and behaviour, since the latter affects feed waste.

Most often, November weight and litter size are included in the breeding goal, receiving approximately the same relative weights. Fur quality is also very important, but the focus on this trait varies more between farms and depends on the farmer's ability and interest in scoring the fur characteristics of his animals. If, on a particular farm, the farmer considers one of the traits especially problematic (e.g. too low litter size), the relative weight of this trait will typically be increased. Farmers typically do not accept a decline in genetic level of any of the three main traits (fur quality, body weight and litter size).

Genetic Evaluation and Parameters

Genetic evaluation of fur animals is conducted within-herd using standard software created for this purpose. For fur animals this works quite well, because the entire breeding programme operates at herd level, and because genetic evaluations, and the assessment of environmental effects, are relatively simple when compared with those for other livestock species such as cattle and horses. Environmental influences on phenotypic performance can be considered homogeneous within herd, year, sex and parity.

Kopenhagen Fur provides a software package (FurFarm) for managing breeding to their members. The FurFarm system handles farmer records (performance results and other information about the animal such as its parents) and performs single-trait genetic evaluations using an animal model including a few environmental effects (sex and year) in addition to random effects of animal and permanent environment.

Along with FurFarm, a few other commercial systems are being marketed. The Morsø Winmink system is the alternative most often used in Denmark. It works in a similar way to the Kopenhagen Fur system, although it does perhaps use more approximate methods for genetic evaluation and mating proposals. A decisive factor for breeders is how flexible and user-friendly the software is; it is also important for the system to produce statistics that assist in the practical management of the population.

We have limited knowledge of genetic correlations. However, unfavourable genetic correlations have been found between weight and pelt quality, as well as between weight and litter size (e.g. Lagerkvist et al., 1994). In practice unfavourable genetic correlations are handled by attaching appropriate relative weights to the relevant traits.

Organization and Breeding Programme

The breeding programmes are organized within-farm. This flat breeding system, which is unique among domesticated animal species, is due to the relatively low reproductive rate of males in fur animals and the rapid changes in demand for different kinds of fur product. As a result of the flat breeding structure any genetic progress achieved at one farm is spread relatively slowly to the rest of the population. Hence, there is also a considerable variation in genetic levels at different farms (about the same genetic variation between farms as within). Also, inbreeding is mainly a problem within herds, and it can usually be alleviated by buying new breeding animals from another herd. However, care should be taken where new breeding animals are always purchased from the same farm, as they may be genetically related to previous breeding animals in this case.

Although breeding programmes are run on a within-farm basis, a limited exchange of animals between farms does exist — e.g. to avoid inbreeding or introduce new lines. However, objective selection across farms is difficult due to the lack of across-herd breeding values. Instead fur farmers rely on customized top lists ('Hit-lists') produced by Kopenhagen Fur (web service). These lists give details of farmers receiving the highest prices per pelt and with superior performance, and of nearby colleagues having the desired type of animals. For several reasons the Hit-list cannot be relied upon to optimize breeding decisions: (1) average prices and other performance parameters are influenced by the environment and other non-genetic factors; (2) the lists do not account for animals being sold for breeding rather than pelting, which disadvantages farms that sell many of their best animals; (3) a farm average is not necessarily indicative of the breeding value of selected (e.g. worst) animals at the farm.

It would seem fairly easy to extend and adapt the Kopenhagen Fur system so that it gives across-farm evaluations — at least, for farms which (a) are genetically linked because, for example, they have exchanged breeding animals, (b) measure traits in a similar way, and (c) can provide unique animal identifications. Of these conditions, it is (b) that presents the main obstacle, since certain traits (e.g. fur quality) may well not be measured in the same way on different farms.

The FurFarm and Morsø Winmink software systems provide mating proposals that help to avoid the mating of closely related animals and hence control inbreeding. However, many farms do not keep proper records of their animals. Some of these use a rotation system to reduce inbreeding, where they use males born in barn 1 in barn 2, males born barn 2 in barn 3, and so forth. This strategy is quite effective in controlling inbreeding when applied systematically.

The selection process can vary between farms, but the number of selected males and females is a function of the expected number of offspring per animal and the desired number of animals for the coming year. Thus, if a farmer has room for 10 000 mink in the coming year (same as current year) and expects 5 surviving pups per Female, he needs to select 2000 breeding females (40% of female pups born corresponding to a selection intensity of 0.97). Likewise he needs to select 400 males (8%, i=1.86) to maintain the population.

The selection is conducted in different steps, and there is usually a significant pre-selection of pups before the final selection in November. This pre-selection is an overall subjective assessment by the farmer. It is usually based on the animal's own phenotype, but in case of pelt gnaw, siblings and the dam are also often discarded. Normally the best females are kept for 2–3 seasons and the worst ~1/3 of the females are culled after their first season. The males are typically culled straight after mating, as they still have their winter fur in late March (i.e. they are only used for one season).

Examples of Genetic Trends

In Denmark the average litter size of mink has increased from 3.6 kits in the early 1970s to 5.5 kits in 2009. Likewise, the November weight has increased considerably. The proportion of 'large' pelts sold at Kopenhagen Fur between 1998 and 2005 increased from 24% to 78% for males, and from 44% to 89% for females. Breeding played a considerable role in bringing about these improvements; however, it is not known precisely how much of the gain was brought about by an improved environment.

Sheep Breeding

The main purpose of sheep farming in Denmark is lamb meat production. Most breeds also produce wool. However, normally the cost of shearing exceeds the market value of the wool in Denmark, although it does not do so in other countries such as Australia and New Zealand. Around the world there is also a significant market for sheep milk — mainly for cheese production and especially in Middle Eastern countries. Worldwide there are about 1.1 billion (i.e. $\times 10^9$) sheep. The population has been fairly constant over the past couple of decades (FAO, 2008). Sheep are hardy animals, well suited to extensive animal production. They are often kept on land, such as mountainous terrain, which is unsuitable for other forms of agriculture. In Denmark sheep are often put to grass around fish farms, conifer trees and on grass seed fields after the harvest.

Denmark is a minor player in global sheep production. There are approximately 3000 small flocks with an average of 20 sheep per flock in Denmark. Only few of these (430 herds) participate in official performance-recording, and most of these recorded flocks are kept on small farms. In all ~8500 lambings are recorded each year in Denmark, but this equates to only 18 lambings (30 lambs) per herd on average. There are only a few big commercial flocks with up to 2400 ewes; these graze permanent grass pastures in the summer, and after weaning the lambs and ewes are allowed to graze on grass seed fields

Both ram and ewe lambs are able to reproduce at 6 months old, but most often the ewes will lamb for the first time as two-year-olds. Most breeds produce 1.5 lambs per ewe per year. The lambs can be weaned at 3 months of age. One ram can mate with up to 100 ewes in a breeding season. Gestation lasts approximately five months.

Breeds

A vast range of breeds and crosses are used around the world: for example, 282 breeds are described at http://www.ansi.okstate.edu/breeds/sheep. Just 25 breeds are represented in Denmark, however, where Texel, Oxford Down, Dorset, Shropshire and Suffolk are the numerically dominant breeds.

Here we focus on breeds used in Denmark for meat production. These can be divided into groups to reflect the special qualities that they contribute to a crossbreeding system. Soundness of mouth, feet and legs, and a dense, protective fleece are of primary importance, together with good body conformation, which is associated with a fat layer, as body reserves for winter survival are important in all breeds.

Ram breeds: These are specialized meat breeds. They have a relatively large growth capacity of up to 0.7 kg a day and can produce slaughter lambs with good carcass quality. The specialized breeds in Denmark are as follows (the number of recorded ewes in 2008 is given in parentheses after the breed name): Texel (1900), Dorset (800), Oxford Down (550) Suffolk (800) and Shropshire (1000). Sheep in the ram sheep group do well on improved and well drained grassland on mineral rich soil. It should be noted that Texel absorb more copper and phosphorous in the intestine from a given feed than all of the other breeds. This is an advantage when the feed contains low levels of these minerals, but it also increases the risk of poisoning and other disorders when the level of these minerals is high.

Dual purpose breeds: These are generally characterized by good mothering abilities, such as easy lambings, sufficient milk yield for raising lambs. They still maintain a relatively good growth capacity and have a medium carcass quality. The following breeds are typical for this category: Leicester (240), Marsh (250), Rygja (100), Dansk Landrace (325) and Såne (150).

Ewe breeds: These are specialized for lambing. Lambing is likely to be easier in these breeds since they have fairly narrowly placed shoulders combined with a large width in the pelvic region; the lambings normally result in very vigorous lambs. Breeds include Gotland Pelt (400), Spel (400), Iceland (70), Finnsheep (120) and Gute. All of these breeds belong to the group Nordic Shorttail sheep. The listed breeds, with the exception of Finnsheep, do well at a low stocking rates on natural vegetation on sandy soils. Finnsheep deviate from the other breeds in the group by having a very high level of fertility, possibly caused by a major gene. Texel×Gotland ewes are popular among commercial lamb producers in Denmark.

Dairy breeds: These are used to increase milk production. The lambs are weaned shortly after lambing and the milk is sold for cheese production. Breeds include Friesian (130) and Lacaune. These breeds are not widely used for crossbreeding, because they do not perform well under extensive conditions. Instead they are used for dairy production in intensive systems, and there are very few such herds in Demark.

Wool breeds: These sheep are used for wool production. In Denmark there are only a few Merino sheep, which have very thin wool fibres, imported from Tasmania.

Genetic Evaluation and Parameters

The Danish Knowledge Centre for Agriculture in Skejby is responsible for calculating breeding values for sheep. For this purpose they use farmer records of lamb mortality (required), lambing ease (voluntary), litter size (required), birth weight (voluntary), weight at 2 months (voluntary) and weight at 4 months (voluntary). In addition to this, ultrasonic scannings of Longissimus Dorsi muscle and fat depth, as well as conformation scores performed by trained technicians, are used. The latter is not widely performed, whereas 10–20% of lambs are scanned for the beef breeds. The small number of records means that the breeding values of most animals have low accuracy. Old carcass records from slaughter houses are also still used, although new records have not been added since 2006, because carcass traits measured at slaughter are genetically correlated with similar traits measured by scanning living animals

Breeding values are estimated using multi-breed Animal Models. Explanatory environmental effects considered vary from trait to trait, but often include interactions of breed with herd×year, sex, season, and age, as well as permanent environment. Both direct and maternal genetic effects are included in the models to account for the effects of both the lambs own, and the mother's, genetic make-up, respectively.

Genetic parameters have recently been estimated by Maxa et al. (2005, 2007, 2008, 2009) and by Norberg et al. (2005, 2006). These estimates, combined with commonly used genetic parameters in sheep breeding, form the basis for the estimation of breeding values. The assumed heritabilities and phenotypic standard deviations. Genetic correlations of 0.8 and 0.7 are assumed between growth rate at 2 and 4 months of age (maternal and direct, respectively), 0.65 between carcass form score and muscle depth, and 0.5 between carcass fat score and fat depth. Furthermore longevity at 1, 3 and 5 years of age are highly correlated genetically (0.86–0.96).

Health traits are not included in the S-index. Examples of genetic impact on health are found in resistance to parasites and scrapie. Heritability for resistance to internal parasites like *Haemonchus* is about 0.3. Resistance to myasis is about 0.25. Resistance to scrapie receives considerable attention in other countries, such as the United

Kingdom, where scrapie is a major problem. A major gene that controls scrapie has been found, and a genetic test is available. The test is not used by Danish sheep farmers, except when it is required for exporting, as Denmark is declared scrapie free. Sheep that are homozygotic for the allele ARR have a very high degree of resistance to the prion which causes the disease. The prion has an effect on the central nervous system similar to that witnessed in BSE in cattle. Heterozygotic sheep with one ARR and no VRC allele are partially resistant to this effect.

Marker assisted selection is not used in Danish sheep breeding, although some knowledge of single genes exists. The Callipygian gene is an example of a single gene that has a markedly positive effect on carcass quality. The presence of this gene results in increased dressing of 5–8%, increased loin eye area of 22–34%, and decreased depth of fat on the back of 25–32%. Despite the general improvement in carcass quality it causes, this gene has a significantly negative effect on meat quality.

Coat colour is a quantitative trait that is important from the standpoint of breed. It is determined by one or a few pairs of genes with a sharp distinction between phenotypes.

Autosomal recessive defects are rarely seen today, because breeding programmes are designed to exclude them. Examples of such defects are inherited blindness in Texels, 'naked lambs' (an inherited disorder of thyroid metabolism) in Dorsets and Merinos, and cleft palate in Shropshires.

Organization and Breeding Programmes

There is no properly organized overall plan in Danish sheep breeding, although there have been several attempts to establish one. From a theoretical standpoint it could be claimed that if breeders use animals with a high S-index for breeding, they are following a de facto breeding plan. However, often breeders prefer to base their selection decisions on their own subjective assessments. The breeders are proud of their stock and sometimes rely on old customs based on intuition. This may work well for highly heritable traits such as growth and muscle depth, but it is less effective for less heritable traits such as litter size, lambing ease, lamb vitality and longevity.

Most breeders avoid mating closely related animals. Due to this, and the fact that much selection emphasis is based on phenotypes rather than breeding values, inbreeding is typically not considered a problem in Danish sheep breeds. Norberg and Sørensen (2006) estimated inbreeding rates for the past decade of 1.0–1.1% for Texel, Oxford Down and Shropshire, which is an acceptable inbreeding rate. Inbreeding may be more of a problem in smaller breeds, but this has not been investigated.

Ultrasonic scanning of the fat and muscle depth of the back muscle is mostly carried out in flocks of ram breeds. Linear conformation assessment is carried out in only a few flocks, and the data are included in the index for body conformation with a low economic weight. This may explain the limited interest in this breeding assessment.

Results from livestock shows are not integrated in the S-index, but these results carry a great deal of prestige for many breeders. Breeders who have obtained top results several times have a high status among colleagues and can use their reputation to sell breeding animals. Some breeders claim that the prizes they are awarded at the shows are the only benefit they get from their breeding work. However, marketing and social engagement are often the main reason for showing animals.

Horse Breeding

Before industrialization the horse was an important provider of pulling power in the agricultural sector. Even in the 1940s there were about 600 000 horses in Denmark, but thereafter the numbers dropped to around 60 000 in the mid-1960s. Since then the horse has enjoyed a renaissance as sports and leisure animal, and today there are about 200 000 horses in Denmark. To put that figure in perspective, there are 59 million horses worldwide (FAO, 2008).

Today horses are used in many different ways. As well as being used for hobby and leisure, horses are ridden or driven in a wide variety of sporting competitions. They are also employed in less traditional areas such as health therapy, tourism and nature preservation. An analysis of the economic influence of the Danish horse sector was produced in 2010 showing a total turnover of 23 369 million DKK, and that 20 849 full time jobs have been created in and by the sector.

Horse breeding in Denmark is organized around the National Committee of Horse Breeding, which consists of representatives of 30 member breeding societies and covers more than 95% of the registered breeding horses. The committee deals with overall political and strategic matters within horse breeding, and seeks to establish common guidelines and rules for breeding and registration. The equine section of the Knowledge Centre for Agriculture keeps the studbooks for the majority of breeding societies represented on the National Committee and maintains a *National Horse Database.*

Breeds

A wide variety of breeds are handled in Danish Horse breeding. They are traditionally divided into the following groups on the basis of shared, or somewhat similar, characteristics:

Special breeds (group A). The group of special breeds consists of lighter, noble and specialized breeds. Several of them, including Frederiksborg, Lipizzaner and P.R.E., are baroque breeds with more than 400 years of history behind them. This group also contains highly specialized racing horses such as Thoroughbred and Trotters and the oldest breed in the world, Arabian Thoroughbred. Other breeds in the group include Oldenburg, Pinto, Friesian, Shagya, sports and Anglo Arabians, Quarter Horse, Paint Horse, and Appaloosa.

Danish Warmblood – National riding horse (group B). The Danish Warmblood has been part of organized horse breeding in Denmark since 1962. Before this, there was no specialized, modern riding horse breeding in Denmark. The primary purpose of these horses is dressage and jumping competitions.

Draft horses (group C). The main goal for these heavy horses is to produce power for draft work. Therefore they need to be powerful horses, with a co-operative and willing temperament. Breeds here include Jutland horses, Belgian horses, Shire, and North Swedish Working Horse.

Smaller horses (group D1). Horses in this group have traditionally performed duties originally believed to be suited to larger horses. The group consists of horses with remarkable 'original traits' providing good fitness in natural environments, a good temperament and high health status. Today the group is mainly used as versatile hobby and sports horses. The breeds include Fjord, Icelandic horses, Haflinger, and Tinker (or Irish Cob).

Ponies (group D2). Pony breeds display broad variation in size and background, from the Danish Sports Pony (DSP), which represents breeding for specialized types of riding pony, to the English Mountain and Moorland breeds, to the Miniature Horses. Common to all is good temperament, making various kinds of use possible, and making these ponies a good starting horse for children. Breeds also include

Connemara, New Forest, Dartmoor, Gotland Russ, Welsh Ponies and Shetland Ponies.

Breeding Goal

We shall focus exclusively on the horse with the largest horse breeding association in Denmark, the Danish Warmblood (DWB); breeding for the other breeds is less developed. DWB is an open studbook that uses many international warmblood breeds to produce riding horses of a specifically defined type and function. It is an advantage of an open studbook that inbreeding is easier to control.

The present breeding objective of DWB is: 'A noble, leggy and supple riding horse with high rideability and a strong health. It has capacity in either jumping or dressage to compete at international level".

Before 2004 the breeding goal of DWB was to produce all-round sport horses that were able to compete in both jumping and dressage competitions. However, in 2004 the studbook initiated a division of the breed into dressage and jumper specializations, because the genetic correlation between the two functional qualities was estimated to be negative (Nielsen & Pless, 2007) and because there was a growing demand among riders for specialized horses. The population now consists, approximately, of 65% dressage-adapted horses and 35% jumping-adapted horses.

With horses it is difficult to estimate economic values for each trait, as many approach horse breeding as a hobby and have no, or very low, expectations of profit; again, the market value of a horse is often influenced by fancy. As a result of this, DWB does not have a total-merit breeding goal of the sort that combines all traits in a single index, although sub-indexes of young horses for jumping and dressage are published. The unavailability of a total-merit index makes focused, systematic breeding difficult and leaves the difficult decision of weighing traits against each other to individual breeders.

In her master's thesis, Mia Haagensen investigated the realized selection emphasis of an 'average' DWB breeder by correlating breeding values of stallions for dressage, jumping and conformation with the subsequent increase in numbers of progeny. This work revealed that breeders put equal selection emphasis (selection index weights) on dressage and conformation, and twice as much emphasis on dressage than they do on jumping. Health and longevity are also part of the breeding goal, but currently breeding values are missing for these traits and there is no direct selection emphasis on them other than natural selection (and some phenotypic selection against health disorders such as osteochondrosis). Typically, a riding horse is 9–11 years old before its performance peak. By then, many years of intensive training have been spent on the horse, so longevity is a crucial quality. DWB has recently started to collaborate with Danish veterinarians on pathological registration. In the future, veterinary diagnoses will be uploaded to the Danish horse database so that the information can be utilized in future selection procedures.

Information on performance traits becomes available late in the horse's life, and efficient selection is further complicated by low heritabilities for those of these traits of main interest. Data on conformation and traits recorded at young horse tests are available earlier, and these traits are generally more heritable. Conformation can be recorded already in foals. Thus, these traits are useful as indicator traits and can help the breeder to achieve correlated genetic progress if they are strongly correlated with performance traits, genetically. Nielsen & Pless (2007) estimated genetic correlations among all traits in the DWB breeding programme using an approximate method. They found, for instance, that the genetic correlation between canter and jumping was moderate to high (0.27 to 0.98). Therefore, canter can be used in early selection for jumping abilities.

The genetic correlations between traits recorded at young horse tests and performance traits are high for both dressage and jumping (~0.8). Genetic correlations between dressage and conformation traits are moderate to high (0.16–0.89). Some conformation traits are therefore well-suited as an indirect selection of dressage performance. Genetic correlations between jumping and conformation traits are also positive, but lower (0.04–0.49). This means that conformation traits are more useful for improving dressage performance, while young horse tests are more important for improving jumping performance.

Dog Breeding

The history of the domestic dog can be traced back at least 15 000 years and possibly as far back as 100 000 years. The earliest archaeological evidence of a domesticated dog is a mandible from a grave at Oberkassel in Germany; it is 14 000 years old. However, archaeological findings tend to underestimate the period of domestication, and an analysis of the mitochondrial DNA of the mandible implies that the origin of the dog is considerably more ancient. Dogs evolved from the grey wolf through various advances in domestication involving repeated genetic exchanges between dog and wolf populations. Domestication has been accompanied by a variety of human needs for assistance with, for instance, herding and hunting. Selective breeding over recent centuries has ensured that dogs now display tremendous variation in their behavioural, physiological, and morphological phenotypes, resulting in over 400 genetically distinct breeds. The dogs exhibit a huge variation in body size — indeed greater variation in this respect than any other terrestrial mammal species. It is not easy to establish an accurate estimate of the total number of dogs worldwide — for instance, the number of stray dogs in Delhi, India is thought to be 200 000. The five countries in the world with the largest dog populations are: USA: 60 million; Brazil: 30 million; China: 22 million; Japan and Russia: 10 million each. In Denmark dog registration is statutorily required. All dogs must be registered in the 'Dansk Hunderegister' (Danish dog registry). The Danish dog population comprises approximately 550 000 dogs of which 60–70% are purebred; so the remaining 40–30% of Danish dogs are crossbred. Sexual maturity in the dog develops at 6–12 months (with the latest arrival occurring in large dogs). Pregnancy is possible in the first oestrus cycle, but breeding is not recommended before the second cycle. The average length of the reproductive cycle for females is 6 months. Dogs bear their litters roughly 9 weeks after fertilization, although the length of gestation can vary from 59 to 65 days, with 63 days being the average. The average litter size is about 6 puppies, but the number varies greatly between breeds. Toy dogs, for instance, produce 1–4 puppies, while larger breeds may average as many as 14 puppies per litter.

Breeds

The first evidence of distinctive breeds of dog dates to 3000 years ago in ancient Egypt. Early Egyptian art illustrated two types of dog: one that was slender with erect ears and a curly tail; and another that was shorter with a heavy muzzle and drop ears. Since then, a broad variety of breeds have been developed, ranging from the diminutive Chihuahua to the giant Irish wolfhound. The more than 400 breeds that are recognized worldwide are traditionally divided into 10 groups according to their morphological or functional characteristics. The 10 groups are also used in dog shows and competitions.

Group 1 Sheepdogs and Cattle-Dogs. Breeds like the German Shepherd, the Belgian Sheepdog, the Collie, the Shetland Sheepdog, the Border Collie and the Old English Sheepdog are included in this group. The original use of these breeds was to help the shepherd when he was gathering his flock together or when he wanted to move the flock to another grazing area. The sheepdogs are agile and alert dogs, and today many of them perform very well in competitions involving agility or obedience.

Group 2 Pinschers, Schnauzers and Molossoid Breeds. Group 2 is more diverse, ranging from small pinchers to the heavy Mastiff types of dog originally used to guard homes and property. The group also includes the guarding shepherd breeds like the Bernese Mountain Dog, the Saint Bernard, the Leonberger, and the Pyrenean Mountain Dog. Popular breeds like Boxers and Newfoundlands, and the national Danish dog breed, the Broholmer, are also in group 2. The Schnauzers come in three sizes: miniature, standard and giant.

Group 3 Terriers. The terriers were originally used for hunting, but today they are mostly companion and family dogs. Most of the terriers have wirehair coat that offers good protection from the elements.

Group 4 Dachshunds. The dachshunds were developed for hunting foxes and badgers out of their earths and setts. They belong to the so called 'chondrodystrophic' breeds. This term is used to characterize the phenotypical shortness of leg caused by an inherited type of dwarfism. Dachshunds come in three hair-variants (smooth-haired, long-haired and wire-haired) and three size variants (standard, miniature and rabbit).

Group 5 Spitz Dogs and Primitive Types. The Nordic spitz dogs, like the Greenland Dog and the Siberian Husky, were originally used to pull sledges. Spitz dogs like the Norwegian Elkhound were also used for hunting. The primitive types include breeds like the Mexican and Peruvian hairless dogs and the Basenji.

Group 6 Scent Hounds and related breeds. The Bloodhound, the English Basset and the Beagle belong here, as do the French Basset types. Norway and Sweden have a number of national breeds in this group as well — for instance, the Hygen, Hamilton and Schiller Hounds and the Swedish Drachsbracke.

Group 7 Pointing Dogs. Pointing dogs are used for hunting. They include breeds like the Pointer, the English and Irish Setters and the

German smooth-haired or wire-haired Pointers. They all have the ability to 'freeze' in a pointing or setting gesture when they identify a bird or other hunt animal. This gives the hunter an opportunity to approach closer before he allows the dog to move forward and flush the bird.

Group 8 Retrievers, Flushing Dogs and Water Dogs. These are hunting dogs as well, but their hunting skills are different from those of pointing dogs. They work in the near vicinity of the hunter and flush the birds immediately they find them. Many of the breeds in this group are also excellent retrievers — i.e. will pick up the dead birds and bring them back to the hunter. Their retrieving ability is especially valuable in duck hunting, where the birds often have to be fetched from water. The group includes breeds like the Labrador and Golden retriever, as well as variants of the Spaniel.

Group 9 Companion and Toy Dogs. This is a large and diverse group including breeds like the Poodle, the Lhasa Apso, the Chihuahua, the Pekingese, the Pug and the Boston Terrier. These breeds are companion dogs, and as such they are bred mainly for their easy, amiable nature and special phenotypic appearance. In this group we find dogs with spectacular coats, and breeds with flat noses, rounded skulls and eyes that to some extent mimic those of a child.

Group 10 Sighthounds. Sighthounds are bred for speed and elegance. The majority of the breeds in this group originate from Great Britain or the Middle East. The group includes the Greyhound, the Whippet, the Saluki, the Afghan Hound, the Deerhound and the Irish wolfhound. The dogs were developed originally for hunting, but today their running speed is capitalized upon mainly in dog racing.

Each breed is registered at the international dog society Fédération Cynologique Internationale (FCI) and has a country of origin — an 'owner' country. The owner country writes the international standards and thus defines how the ideal dog should be in respect of various phenotypical features, such as type, health status and behaviour. The standards are applied by judges at dog shows. Denmark is the owner country of five breeds: the Greenland Dog, the Broholmer, the Danish-Swedish Farm Dog (together with Sweden), the Danish Spitz (not yet approved by the FCI) and Old Danish Pointing Dog.

The most popular breeds in Denmark are the Labrador Retriever, the German Shepherd and the Golden retriever.

Bibliography

Adams, C.E. : *Mammalian Egg Transfer*, Boca Raton, FL, CRC Press, 1982.

Adams, Carol J.: *Animals and Women: Feminist Theoretical Explorations.* Durham, NC: Duke University Press, 1995.

Archana Satarkar: *Food Science and Nutrition*, ABD Pub, Delhi, 2008.

Arora, N. : *Manual of Animal Nutrition*, International Book, 2004.

Arseniev, V.A.: *Atlas of Marine Mammals*, Neptune City: T. F. H. Publishing, Inc., 1986.

Arti Sharma: *Fishes : Aid to Collection, Preservation and Identification*, Daya, Delhi, 2006.

Aruna T. Kumar : *Handbook of Animal Husbandry*, Indian Council of Agricultural Research, 2008.

Baker, Steve: *The Postmodern Animal.* London: Reaktion Books, 2000.

Balram Pani: *Textbook of Animal Chemistry*, I K International, Delhi, 2007.

Basavaraj S. Benni: *Dairy Co-operative Management and Practice*, Rawat, Delhi, 2005.

Baudrillard: *The Animals: Territory and Metamorphoses. Simulacra and Simulation.* Ann Arbor: University of Michigan Press, 1994.

Bekoff, Marc: *Strolling with Our Kin: Speaking For and Respecting Voiceless Animals.* New York: Lantern Books, 2000.

Betteridge, K.J. : *Embryo Transfer in Farm Animals,* Ottawa, Agriculture Canada, 1977.

Billingham, R.E., and W.K. Silvers. : *Transplantation of Tissues and Cells,* Wistar Inst. Press, Philadelphia, 1961.

Bourbon, Richard M.: *Understanding Animal Breeding*, Prentice-Hall, 2000.

Bower, B.: *Fossils may Clarify Mammal Evolution*, Science News, 1984.

Brock, J. : *A Natural History of Domesticated Animals*, Cambridge Univ. Pr., New York, 1999.

Bronson, F. H.: *Mammalian Reproductive Biology*, Univ. Chicago Pr., Chicago, 1990.

Brown, L.: *Cruelty to Animals: The Moral Debt.* London: MacMillan, 1988.

Brown, R. E.: *Social Odours in Animals Reproduction*, Clarendon Press, Oxford, 1985.

Bushnell, R.B. . : *Dry Cow Feeding and Management*, A Western Regional Extension Publication, 1979.

Carroll, R. L.: *Vertebrate Paleontology and Evolution*. W. H. Freeman and Co., New York, 1988.

Clark, Stephen: *The Moral Status of Animals*. Oxford: Oxford University Press, 1977.

Clutton Brock Juliet : *Horse power: a history of the horse and donkey in human societies*, National history Museum publications, London 1992.

Clymer, R. : *Nature's Healing Agents,* PA, U.S.A: Dorrance Co., 1963.

Crawford A. : *Experiments and Observations on Animal Heat,* London: Printed for J. Johnson; 1788.

Daniel, J.C. Jr. : *Methods in Mammalian Reproduction*, Orlando, FL, Academic Press, 1978.

Davis, A. : *Let's Eat Right to Keep Fit,* New York, U.S.A: Harcourt Brace Jovanovich, Inc., 1970.

Degen, A. A. : *Ecophysiology of Small Desert Mammals*, Springer, New York, 1997.

DeGrazia, David: *Animals Rights: A Very Short Introduction*. Oxford: Oxford University Press, 2002.

Devender Pratap Singh : *A Handbook of Beekeeping*, Agrobios, 2006.

Devyani Khemka: *Animal Physiology*, Dominant, Delhi, 2003.

Eisenberg, John E.: *The Animal Radiations*, The University of Chicago Press, 1981.

Ensminger, M.E. : *Dairy Cattle Science,* The Interstate Printers & Publishers, Inc., Danville, 1980.

Escobar, Roberto Calle: *Animal Breeding and Production of Camelids*, Lima, Peru, 1984.

Flowerdew, J. R. : *Animals: Their Reproductive Biology and Population Ecology*, Cambridge Univ. Pr., New York, 1987.

Gay, W.I. : *Methods of Animal Experimentation*, Academic Press, New York, 1965.

Godthelp, *Animal: Riversleigh. The Story of Animals Reproduction in Ancient Rainforests*, Reed Books, Balgowhah, 1991.

Goel, A K : *Basic Concept of Animal Chemistry*, Pearl Books, Delhi, 2008.

Gordon, G. A.   : *Animals Physiology,* Harper and Row, New York, 1989.

Gray, J.: *Animal Locomotion*, Norton, New York. 1968.

Greene, H. W.: *Mode of Reproduction in Lizards and Snakes of the Gomez Farias Region*, Tamaulipas, Mexico. Copeia, 1970.

Griffin, D.R.: *Animal Minds*, University of Chicago Press. Chicago, 1992.

Grzimek, B.: *Grzimek's Animal Life Encyclopedia*, McGraw Hill, New York, 1989.

Hacker, J.B. : *Nutritional Limits to Animal Production from Pasture,* Farnham Royal: CAB, 1981.

Hagedorn, A.L.: *Animal Breeding*, Crosby Lockwood, 1950.

Harrison, R. J.: *Functional Anatomy of Marine Animals*, New York: Academic Press, 1974.

Hulbert AJ, Else PL. : *Mechanisms Underlying the Cost of Living in Animals,* Annu Rev Physiol. 2000.

Hutt, Frederick B.: *Genetics for Dog Breeders*, Freeman & Company, 1979.

Joysey, K. A. : *Development, Function and Evolution of Animal Teeth*, Academic Pr., New York, 1978.

Krieger, Maggie and Richard: *Secrets of the Andean Alpaca - The Field Guide*, Saltspring Island Llamas and Alpacas, 1994.

Lance, J.W. : *Migraine and Other Headaches,* New York, U.S.A: Scribner, 1986.

Lata Bhattacharya: *Animal Biochemistry*, Discovery, Delhi, 2010.

Lyster, S. : *Animals and Their Moral Standing.* London : Routledge, 1997.

Marshall, R.B. : *Breeding Farm Animals*, Asiatic Pub, Delhi, 2006.

Martin, A.M. : *Fisheries Processing : Biotechnological Applications*, Chapman and Hall, Delhi, 2009.

Mathialagan, P : *Textbook of Animal Husbandry and Livestock Extension,* International Book Distributing Co, Delhi, 2005.

Matthews, L. H.: *The Life of Animals Reproduction*, London, Weidenfield and Nicholson, 1969.

Mindell, E. : *Mindell's Vitamin Bible*, New York, U.S.A: Warner Books, 1980.

Montgomery, G. G.: *The Early Placental Mammal Radiation Using Bayesian Phylogenetics*, Science, December 2001.

Muybridge, E. : *Muybridge's Complete Human and Animal Locomotion*, Dover Publ., New York, 1979.

Nyholt, D.H. : *The Vitamin & Herb Guide,* Alberta, Canada: Global Health Ltd., 1992.

Rathnakumar, K : *Fish Processing Technology and Product Development*, Narendra Pub, Delhi, 2008.

Raymond, F., Redman, P., & Waltham, R. : *Forage conservation and Feeding.* Ipswich: Farming Press, 1986.

Renaville, R and A Burny : *Biotechnology in Animal Husbandry*, Springer Pub, 2008.

Rhykerd Charles L. : *The Cycles of Plant and Animal Nutrition*, Scientific American Books, San Francisco 1976.

Robinson, Roy: *Genetics for Dog Breeders*, Pergamon Press, 1990.

Safley, Michael: *Synthesis of a Miracle*, Northwest Alpacas, 2002.

Schiller, A. L. : *Anatomy of the Guinea Pig*, Harvard Univ. Pr., Cambridge, 1975.

Seidel, S.M.: *New Technologies in Animal Breeding*, Orlando, FL, Academic Press, 1981.

Shagufta Jamal and H P S Arya : *Participatory Rural Appraisal in Agriculture and Animal Husbandry : A Training Manual*, Concept, 2004.

Short, R. V.: *Reproduction in Mammals*, Cambridge, Cambridge University Press, 1972.

Shukla, M K : *Brain Teasers : Multiple Choice Questions on Animal Husbandry and Veterinary Sciences*, International Book Dist, Delhi, 2007.

Singh,. G : *Chemistry of Amino-Acids and Proteins*, Discovery, Delhi, 2007.

Stuart Patton : *Principles of Dairy Chemistry*, Huntington, N.Y.: Krieger, 1976.

Thornhill, Nancy W.: *The Natural History of Inbreeding and Outbreeding*, Chicago Press, 1993.

Verma, S.R. : *Nature: Fish Genetics and Biodiversity Conservation*, Conservators, Delhi, 1998.

White, M.J.D.: *Animal Cytology and Evolution*, Cambridge, Cambridge Univ. Press, 1954.

Yablokov, A.: *Variability of Mammals: Moscow*, USSR, Nauka Publishers, 1966.

Yadav, Manju: *Mammalian Development*, Discovery Publishing House, Delhi, 2008.

Index

◻◻◻